Video Camera Techniques

Gerald
Millerson

FOCAL PRESS
London & Boston

Focal Press
is an imprint of the Butterworth Group
which has principal offices in
London, Boston, Durban, Singapore, Sydney, Toronto, Wellington

First published 1983

British Library Cataloguing in Publication Data

Millerson, Gerald
Video camera techniques.
1. Video tape recorders and recording
I. Title
778.59'9'028 TK6655.V5
ISBN 0-240-51225-1

Library of Congress Cataloging in Publication Data

Millerson, Gerald.
Video camera techniques.
(Media manuals)
Bibliography: p.
1. Television cameras. 2. Cinematography
I. Title. II. Series
TR882.M544 1983 778.59 83-14166
ISBN 0-240-51225-1

Photoset by Butterworths Litho Preparation Department
Printed and Bound in England by A. Wheaton & Co. Ltd., Exeter

Contents

Introduction

As technology transformed the bulky broadcast television camera into the compact 'go-anywhere' units we know today, the exciting world of non-broadcast (closed-circuit) video has developed. Video cameras have become a regular communication tool in education, business and industry, and for many other purposes have replaced the film camera. Attached to the home video recorder, the video camera provides an instant-replay memory bank of those special personal occasions that are the fabric of our lives.

Although today's video camera is simple to handle, automated, robust and reliable, do not let that fool you into believing that camerawork has now become a 'point-and-focus' routine. Skilful, discerning operation is as important as ever if you want to achieve smooth, persuasive results, whether you are using top-range broadcast facilities or budget video.

In this book you will find a step-by-step study of the techniques and potentials of good camerawork. These are the foundation stones of effective visual communication, whether you are a cameraman/producer shooting your own program material, or one of a camera crew in a multi-camera studio production.

When this book was first published (as *TV Camera Operation*), emphasis was on studio production methods, for camera mobility was limited. This new edition has been extensively revised to reflect changes in camera design and production techniques, and to include the particular problems of the hand-held camera.

Acknowledgements
I would like to thank the Director of Engineering of the British Broadcasting Corporation for permission to publish the original book. I also want to express my appreciation of the encouragement of various colleagues, including Mr C.R. Longman (Controller, Engineering and Operations, BBC Television) and Mr Bob Warman, who, with his many years experience of TV camerawork, very kindly appraised that manuscript.

Remember, the expert had to learn these principles not so long ago!

Meet your Camera

There is something daunting about meeting most technical equipment for the first time. Its impersonal unfamiliarity leaves us at a disadvantage, feeling slightly foolish, perhaps, at not knowing exactly how to handle it. Even when we are accustomed to similar devices, different design or layout can leave us apprehensive. These are natural initial stages. They soon pass.

The video camera is no exception. Remember, your camera is a tool that is surprisingly simple to use once you have grown accustomed to it. Like all good tools, it will repay the trouble you take to find out how to use it effectively.

The video camera is easy to use

You don't need to be 'technically-minded' to use a video camera successfully these days. Although the technology is very sophisticated, it is extremely straightforward to handle. Think of your camera as a communication tool. Like the telephone and the typewriter, it is a device for conveying your thoughts and interpretations to others. It will not do this automatically, simply by being pointed at the scene. You have to use it selectively and with care.

The camera itself is simply an electronic device that continuously produces pictures of the scene in front of the lens, as long as it is powered. No film or processing is involved. Its pictures can be viewed at the very moment they are being shot, or videotaped for replay later.

Successful picture-making

Good cameracraft comes from a blend of skills: easy familiarity with the camera's controls, so that you operate them confidently and accurately; an eye for pictorial opportunities; selecting a good viewpoint, composing effective pictures. Some even less tangible qualities come in useful at times, including dexterity, stamina, patience and a good memory!

Production techniques today take several different forms: the individual program maker operating his own camera; the director with a cameraman and sound-man/assistant; the production team using multi-camera shooting techniques. Each productional approach has its particular opportunities and limitations with a different emphasis in cameracraft.

NOTE: For simplicity, masculine terms (e.g. he, cameraman) have been used throughout this book as a general indication, rather than dual-sex terms such as 'he-she', 'cameraperson' etc.

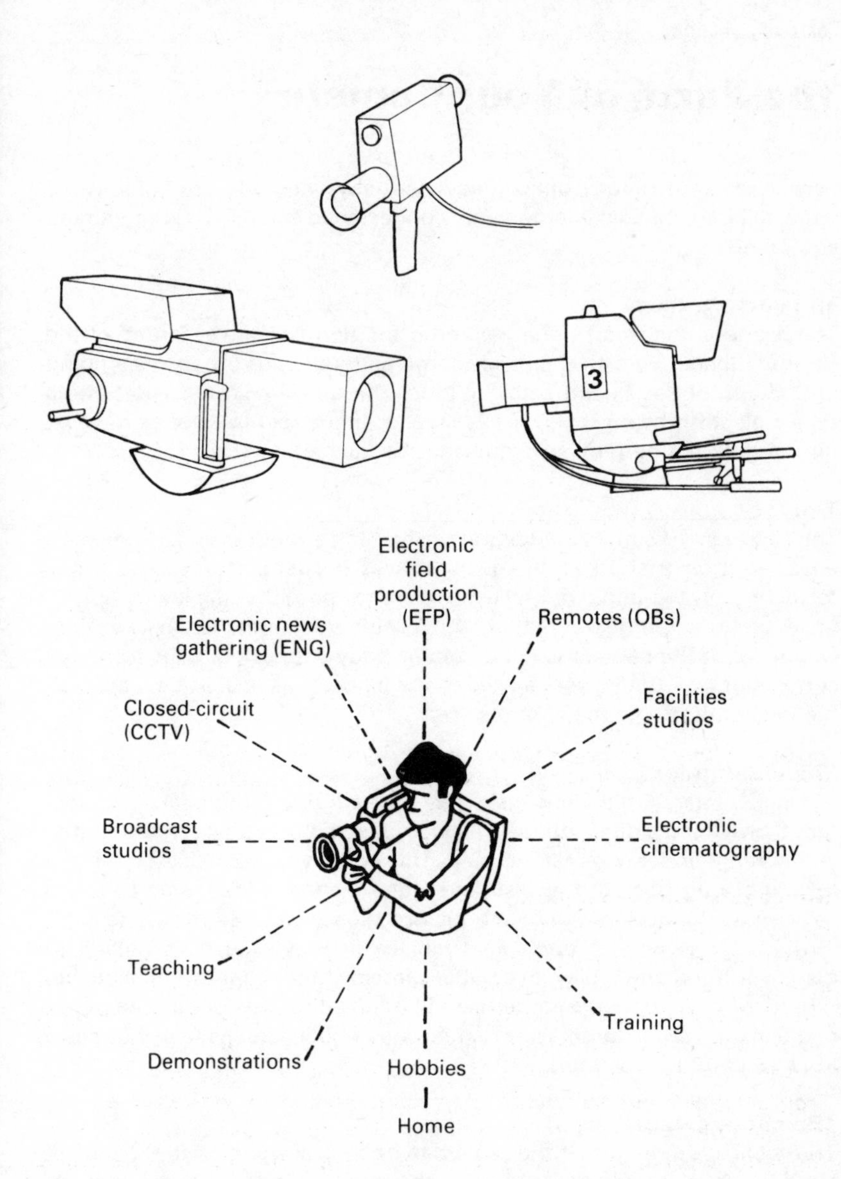

The video camera
Camera design varies considerably: with its price bracket, its intended use, ruggedness and design sophistication required.

A range of shooting conditions
The versatility of video cameras makes them adaptable to all forms of program-making.

The Parts of Your Camera

Camera designs range from compact, virtually self-contained home video units, to large studio cameras, cable-connected to extensive rack-mounted equipment.

The camera tube

The scene is focused by the lens onto the light-sensitive 'target' of the camera tube(s). Here an electrical charge-pattern builds up, corresponding in strength at each point with the tones there in the scene. The camera tube's electron beam rhythmically sweeps this target in a series of lines, discharging the pattern, so producing the fluctuating video signal.

The lens

The cameraman's main interests are in his *lens's* controls and its performance: controls that adjust image sharpness (*focus*) and size (*lens angle/focal length*); the minimum focusing distance possible; the lens's 'speed' (its maximum aperture or *f*-stop); and any inherent limitation in its design. Of course, if the camera is hand-held or body-supported, he is also very concerned with the camera's weight and balance, as well as how accessible and comfortable the controls are.

The viewfinder

In small cameras, the viewfinder may be a simple through-lens arrangement, providing a direct *optical* image as in a photographic camera – often with a rangefinder to assist focusing. This system shows what the camera lens is seeing but not the resulting video picture. Most cameras use a small magnified black-and-white picture tube in the viewfinder (e.g. 40 mm/1.5 in) viewed through an eyepiece. It may also be switched to display a videotape replay. In smaller cameras this viewfinder is fixed, but otherwise it can be repositioned on top or on either side of the camera, as convenient. Larger cameras have open viewfinder screens (e.g. 12.5 cm/5 in to 18 cm/7 in diameter).

The camera cable

The cable interconnecting the camera to its control equipment (CCU; CPU) supplies voltages and waveforms to the camera, and routes its video for amplification and distribution. It may have multi-wire/multi-core cable (e.g. 100–300 m maximum), coaxial cable (e.g. 700–1400 m maximum), triaxial cable (e.g. 700–1200 m maximum), or glass-fiber strand construction (fiber-optics). The last is lightest, interference-free and robust, allowing considerable distances (e.g. over 4000 m) between a camera and its control equipment.

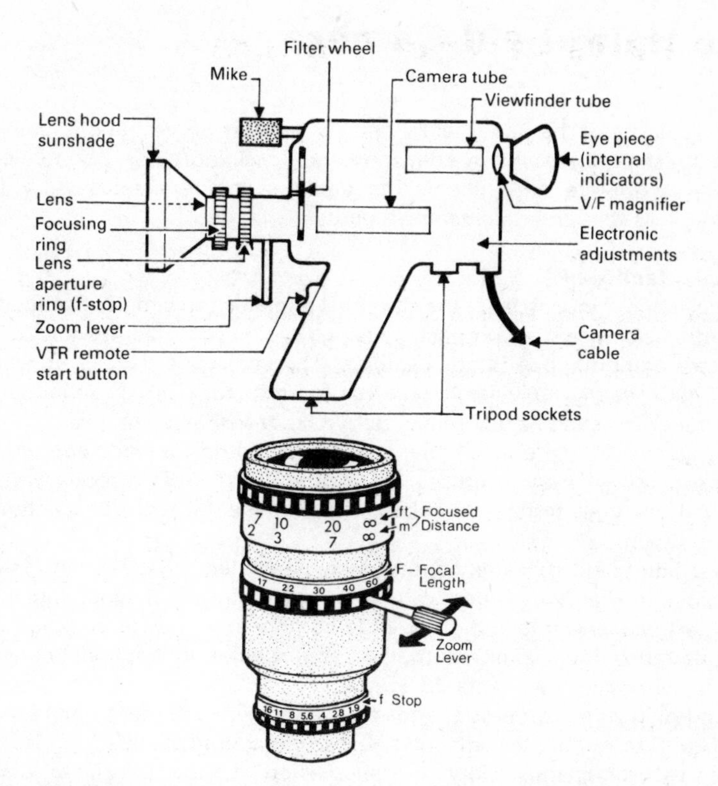

Basic adjustments

Electronic focus: focuses the camera tube's scanning beam for sharpest picture detail.

Beam current: controls scanning-beam strength, set to just prevent lightest picture tones from blocking-off to an even white area. Too much beam current degrades picture clarity and gives excess picture noise. With too little beam, no picture is generated.

Target volts: set for best tonal contrast. If set too low, the camera needs more light to compensate for weaker picture but image lag (trailing) is less. In cameras with vidicon tubes, target volts are adjusted to control exposure, manually or automatically.

Auto-iris: automatic exposure adjustment. The lens aperture (f-stop) self-adjusts to suit the average light conditions of the shot.

Auto-white/white balance/color balance: adjusts the proportions of red, green, blue in the picture to suit the color temperature of prevailing light. Pre-set by pointing camera at a white surface and operating control. (Many cameras have manually selected compensation.)

Auto-black: adjusts picture blacks automatically to a datum black level (lens cap-up).

Zoom switch: rocker-switch controlling zoom in/out.

Exposure/signal level indicator: shows when picture is correctly exposed.

Sensitivity/video gain/boost: video amplification compensating for low light intensities.

Black stretch/gamma correction: improves tonal gradation in shadow areas.

Cable compensation: electronically compensates for definition losses (high frequency fall-off) with cable length.

The Hand-held Camera

The popularity of the hand-held video camera has grown rapidly both in the domestic market and in educational and training fields. Economical, simple to operate and reliable, its pictures are very satisfactory for general, less-demanding closed-circuit use.

Camera facilities

Thanks to ingenious design, the camera's miniaturized simplified circuitry is centralized within a compact housing, avoiding the somewhat bulky ancillary equipment of larger cameras. The camera is usually attached direct to a nearby television receiver (or monitor), or to a videotape recorder (VTR) such as a U-matic, Betamax, or VHS system.

The camera may produce either a standard *composite video output* (i.e. picture video plus sync pulses) or a *modulated radio frequency (RF)* output from an internal 'modulator', to suit the TV receiver's antenna (aerial) input.

Most hand-held cameras contain a single camera tube (17 mm/⅔ in; 25 mm/1 in diameter) fitted with a colored-stripe filter, enabling it to analyze the scene into red-green-blue components. Power supplies are from batteries (internal or external) or from regular utilities/mains power outlets, converted by a suitable adapter.

Hand-held video cameras are normally fitted with a trigger switch on the hand grip, to enable the attached VTR to be started/stopped or paused during recording, thus allowing a succession of shots to appear joined during replay (i.e. 'edited in camera').

A microphone is usually built into the camera, fixed to its top or to a telescopic rod. Although this is a convenient way of picking up the sound, quality is often unsatisfactory compared with a properly positioned separate microphone. It is too likely to pick up extraneous noises.

Synchronizing

All television systems need special *synchronizing (sync)* pulses to ensure that the entire picture channel from camera tube to picture tube is scanning exactly in step. Otherwise pictures would tear, slip, or simply become indecipherable. Hand-held cameras may derive these sync pulses from an internal generator or use those from an attached VTR.

Some small cameras of this type cannot be used in a multi-camera hookup, as their individually generated pulses would not be running in unison. Others use a communal 'sync-pulse generator', or can be precisely interlocked (slaved) by a *genlock* system that coordinates their separate circuits.

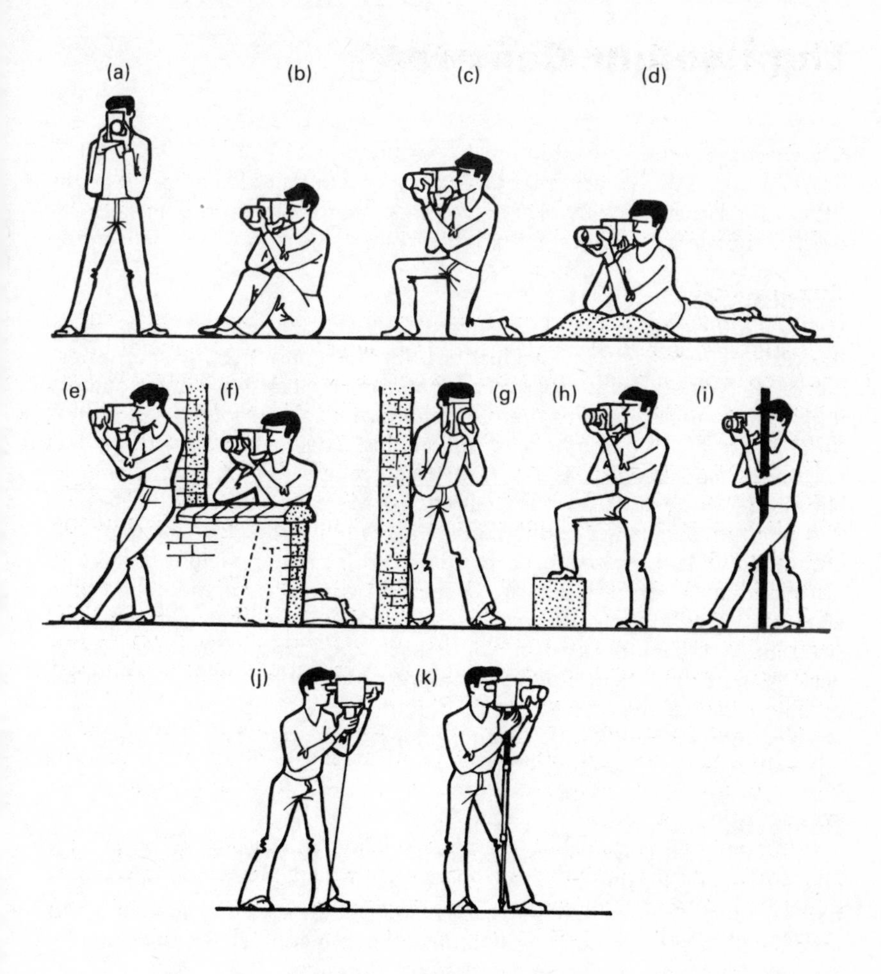

Hand-held supports

Any camera shake causes the picture to weave and hop about, so it is essential to hold it as rigidly as possible. Grip your camera firmly (but not too tightly) with your eye against the eyepiece and both arms tucked well in. Various techniques help to steady the camera:

1. Stable body positions: (a) legs braced apart; (b) seated, elbow on knees; (c) kneeling; (d) ground support.
2. Nearby supports: (e) back to wall; (f) resting on low wall, fence, railings, car etc.; (g) leaning side against wall; (h) foot on step or box; (i) resting against post.
3. String or chain support: (j) attached beneath camera, trapped under one foot and pulled upwards.
4. Monopod: (k) single-leg telescopic tube, or pole.

Freedom to shoot anywhere.

Lightweight Cameras

Shoulder-mounted, lightweight cameras are typically 3.5–13.5 kg/8–30 lb in weight, so a body harness, brace, chest-pad or belt attachment is often fitted to improve stability. This gives the cameraman considerable freedom of movement, even in crowded surroundings or on uneven ground.

Applications

Lightweight cameras are used extensively for video production on location (EFP, electronic field production), for documentaries, drama, interviews, etc. When covering sports events (ESG, electronic sports gathering), their mobility enables them to follow widespread action, e.g. on the touchline at a ball game or on a golf course. They can shoot from a car, boat or helicopter with ease. They are also valuable when improvising or shooting 'off-the-cuff', as when covering news events (ENG).

Apart from specially designed units that combine the video camera with a compact videotape recorder (combos, VRC) the lightweight camera is normally cable-connected to its associated control equipment and to a separate VTR. This may be conveniently carried on a small cart or trolley pack, or in a back pack or shoulder pack, the cameraman often being assisted by someone operating sound, recording and lighting equipment. In some cases, the camera may be routed to a portable low-power transmitter (window unit) which radiates to a nearby relay point or support vehicle.

Convertible cameras

Thanks to miniaturization, the camera tubes, together with a beam-splitter prism (which optically splits the single lens image into two or more light paths for the respective camera tubes) and all the necessary scanning and amplification circuits, can be contained within a compact housing that is suitable for several different camera configurations.

For *field* work, in a shoulder-mount, it can have a small prime lens or zoom fitted and an eye-piece viewfinder attached. In the *studio,* a high-quality optical system with a large open viewfinder can be fitted to the same video package.

Electronic cinematographic camera (ECC)

This is a design concept that provides the *film* cameraman with a video camera which has the facilities and handling characteristics of a 35 mm film camera. Its electronic features provide a performance akin to that of film (adjustable gamma, knee compression of highlights) and typical film-camera systems (prime lenses, matte box, separate 'focus-puller' viewfinder), as well as normal video-camera utilities (video-level meter, VTR time-lapse indicator, auto balance, etc.).

Typical arrangements
Because lightweight cameras can be body supported or fixed to a mounting they offer considerable mobility.
1. Shoulder-mounted, using battery-belt power.
2. Cabled to a back-pack video recorder (or transmitter).
3. A body-brace support, cabled to trolley-pack with video recorder, audio equipment and picture monitor.
4. Cabled to a shoulder-pack VTR.

The Studio Camera

It can be confusing to find that, in practice, the large heavy-duty units called 'studio cameras' are also widely used on location (remotes/outside broadcasts) for such large-scale productions as public events, sports etc., while lightweight and convertible cameras are used both in studios and in the field.

Design features

Studio cameras generally represent the highest 'state-of-art' standards, with their advanced optical systems and electronics. Picture quality is excellent, resolution of detail being higher than that reproduced by regular TV systems (NTSC, PAL, SECAM).

The large *zoom lens* (page 44) is a bulky package designed for maximum performance, and may incorporate an *'extender lens'* for a greater range of lens angles (1.5, 2, or 3 times). Any *filters* (page 106) are usually available in a segmented disc within the camera head – although front-fittings are also used. A precision *prism* assembly (or *dichroic mirrors*) splits the lens image into its three component colors. The camera tubes are typically forms of *plumbicon* or *saticon.*

The camera head, with its large lens and viewfinder systems (up to 18 cm/7 in), may also have a *prompter* device attached (page 132), so a substantial mounting is required. Beneath the camera head a *wedge mount* enables it to be slid onto the *panning head* at the top of a heavy-duty tripod or pedestal. One or two *pan bars (panning handles),* clamped to the panning head, turn it to left and right (pan) or tilt the camera up and down. They can be angled to suit the cameraman.

Controls

In a *lightweight camera* the main adjustments – focus, lens aperture (*f*-stop), zoom – are made on the lens barrel itself by turning it, or sleeve rings, or an attached knob. On the large *studio camera* these functions are adjusted by separate remote controls fixed to the panning handles and/or built into the camera head. Several different designs are popular.

Line-up is the process of adjusting circuitry to ensure accurate color balance, registration, geometry, overall focus, etc. This can be done manually, while shooting a test-card, or automatically by push-button, through computer-controlled adjustments.

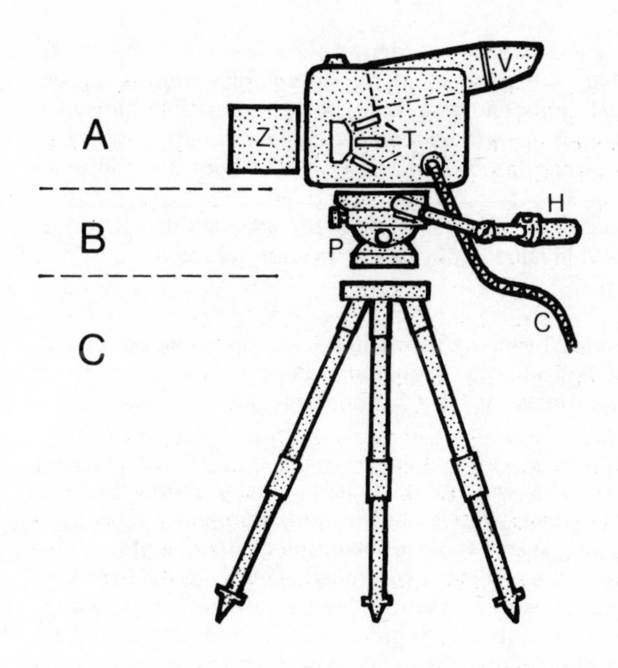

Basic parts of the camera

A. *The camera head* with (Z) zoom lens, (T) camera tubes, (V) viewfinder, (C) camera cable.

B. *The panning head (pan head) enables the camera head to tilt* and turn (*pan*) smoothly. These movements can be controlled by *drag* (*friction*) adjusters or *locked-off* to hold the head rigid. A *tilt-balance* adjustment may prevent the head becoming nose or tail heavy. *Pan bars* (*panning handles*) fitted to the panning head enable the cameraman to direct and control the camera head. Focus and zoom controls are often attached to the panning handles.

C. *The camera mounting* can take several forms, including *monopod, tripod, rolling tripod, pedestal and crane.*

The type of mounting you use determines the camera's versatility.

Camera Mountings

There are various types of camera mounting or dolly. The simplest offer only limited movement, while the most elaborate can move around, raise/lower ('elevate/depress'), swoop to exactly the right position, holding the camera firmly there to pan and tilt as needed.

The tripod

This stationary, highly adaptable three-legged portable support has many applications (page 140). Although standard photographic tripods can be used with the lightest video cameras, many are unacceptably flimsy. A tripod-mounted camera is likely to be top-heavy, so the tripod's legs must be fully spread and firmly based. A support base or floor ties can help here.

Although only able to tilt and pan (height is pre-adjusted) the tripod is easily transported, and invaluable for use in awkward places.

The rolling tripod

When fitted to a castered base, the resulting rolling tripod can be moved around on a smooth level floor. (Uneven floors cause picture judder; floor slope leads to camera runaway). Height is still pre-set.

The pedestal

The pedestal (ped) is the workhorse of large studio production. Its adjustable column has a heavy three-wheeled base, normally guided by a *steering ring* which is also used to raise/lower the camera height.

Depending on design, the column is controlled pneumatically, hydraulically, counterbalanced, or hand-cranked. Pedestals need vertical rebalancing when the camera weight is changed – e.g. a prompter added or removed. Small lead weights are placed in a tray at the top of the column for fine adjustment. Without proper vertical balancing, the column is hard to raise or lower.

Lightweight pedestals provide considerable shot flexibility, rapid height changes, and precision steering, even in confined spaces.

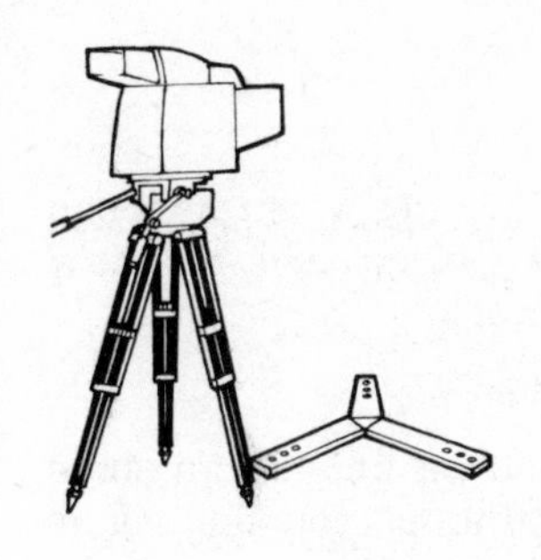

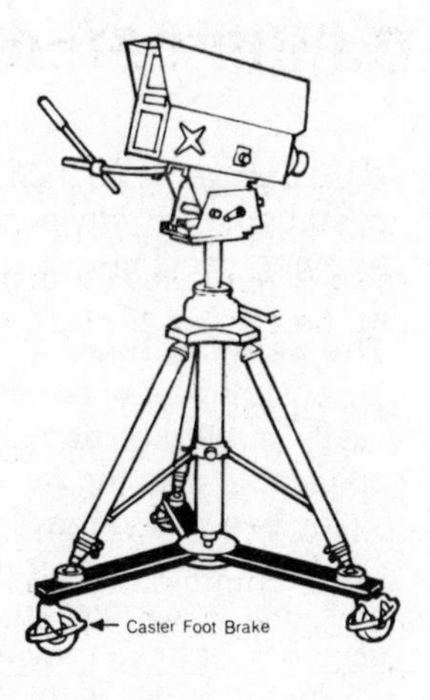

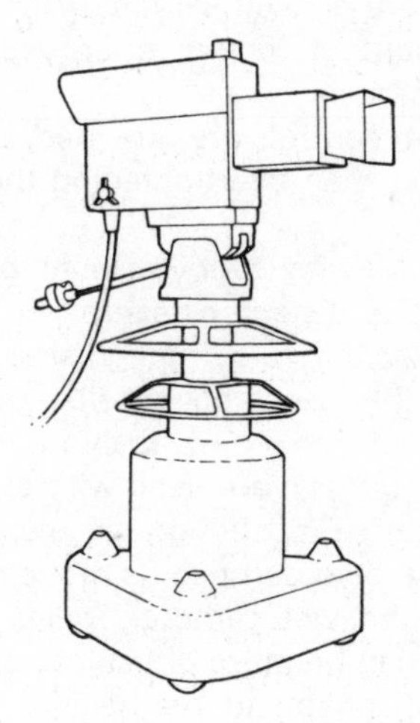

The tripod
A simple three-legged stand, its legs can be extended independently to suit uneven ground. The camera position is fixed but the tripod is portable, compact, easily set up and packed away. A triangular base may be fitted (*spreader, spider*) to improve stability.

Rolling tripod
A heavy-duty tubular metal tripod on a wheeled-castered base. It can be moved around easily on a flat surface but its height is pre-fixed.

Pedestal
Its central column can be moved up/down while on shot or can hold the camera firmly at a chosen height. The top steering wheel guides the mounting and is used for raising/lowering the camera. A bottom ring (if fitted) locks off the column movement.

Camera Movements

Although an audience may accept unsteady pictures when shooting conditions are obviously difficult (e.g. from a moving vehicle), shaky weaving shots are normally very disturbing to watch.

The panning head

To ensure that a camera is firmly supported and yet free to turn about easily in all directions, the camera head is fixed to its mounting via an adjustable panning head/pan head.

The basic head-movements are *tilt* (vertical) and *pan* (horizontal). These can be made easier or more difficult, with drag/friction controls. You will find that a certain amount of friction improves the smoothness and accuracy of head movements.

Where head-lock (brake) controls are provided, use these to prevent tilt or pan movements, rather than overtightening the drag controls, which could ruin them.

Head-locks are very useful whenever you want to keep a shot absolutely still: e.g. on a graphic; a distant close-up on a very narrow-angle/long-focus lens; in a multi-camera combined shot; or to secure the head when leaving the camera (this prevents possible overbalancing); or when moving the mounting around to a new position.

Several forms of panning head are used with video cameras. The *fluid head* is used for lighter cameras, offering more subtle, smoother camera control as layers of silicon fluid dampen its movement. The *friction head,* which is often used for heavier cameras, relies on surface friction to control action. This type can be more difficult to operate smoothly at the start of a movement, and is liable to overbalance when tilted to extreme angles. *Cradle heads,* which are often fitted for the heaviest cameras, maintain good balance at all degrees of tilt.

Moving the mounting

Each type of camera dolly or movable mounting has its own operational quirks. To move dollies around quickly and safely, without juddering the picture or injuring yourself or others, is a technique to be learned and practised; we shall look at this more closely later (page 92).

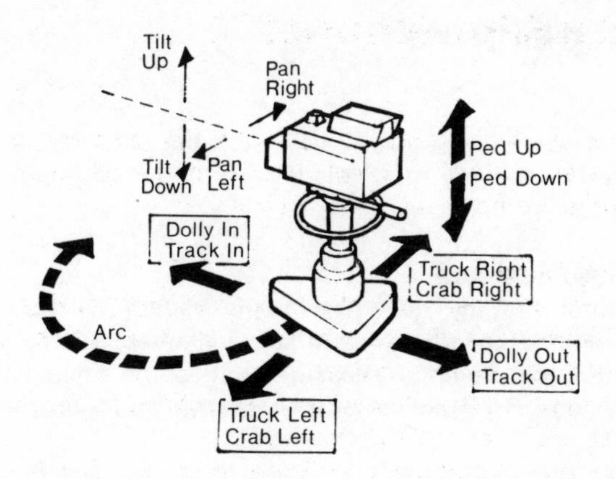

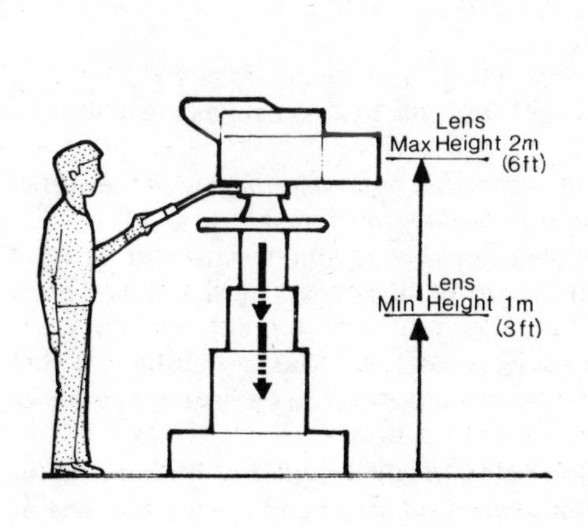

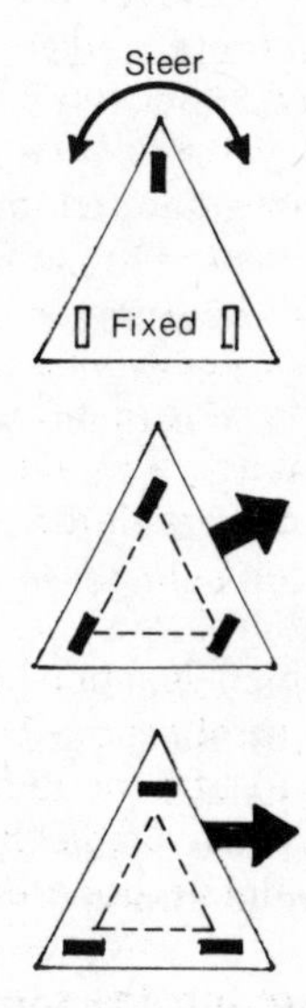

Movements of the camera head
Tilt (up/down) and pan (turn left/right). The mounting can move forwards-backwards (*dolly* or *track* in/out) and move sideways (*truck* or *crab* left/right), and the camera can move up or down (*elevate/depress, ped up/ped down*).

Lens height
Typical lens-height adjustments for a pedestal are from 2m (6ft) maximum to 1m (3ft) minimum.

Pedestal steering
The pedestal's three rubber-tired wheels may be linked to steer together (foot-pedal selection) for sideways moves in a *parallel/crab mode*, or with single-wheel steering (two wheels fixed) in *dolly/tricycle mode* for general dolly moves.

The choice of lens aperture can considerably affect picture impact.

Lens Aperture

Inside your camera lens you will see a circular diaphragm, formed by a series of flat interleaving metal blades. Turning a ring on the lens barrel adjusts the size of this hole.

Adjusting the aperture
This adjustment of the lens aperture is usually calibrated in *f-stops* (calculated values based on its effective diameter) or in *transmission values* (T-numbers) based on measurements of the actual light passed by that lens system. For practical purposes, you can regard these terms as identical.

Whichever method is used for your lens, you will find a series of numbers marked around a lens-barrel ring; a small line or arrow indicates its setting.

The effect of adjustment
Adjusting the lens aperture has two simultaneous effects:
1. It determines how much light from the scene reaches the camera tube (i.e. *'exposure'*, page 26).
2. It affects the depth of the scene that appears sharp in the picture (i.e. *depth of field* – often wrongly called *depth of focus*, page 28).

As you open up the lens to a larger stop (*smaller* number – e.g. *f*/2), more light passes through the lens: increased exposure, reduced depth of field.

As you close down the lens to a smaller stop (*larger* number – e.g. *f*/22), less light is transmitted through the lens: reduced exposure, greater depth of field.

You will usually set the lens *f*-stop to suit prevailing lighting conditions. Under strong sunlight, for example, you may need to stop the lens down (e.g. to *f*/16), while in dim light you will have to open it up (e.g. to *f*/2).

At other times you may decide that you want your picture to have a certain amount of depth. You then choose an appropriate *f*-stop and adjust the light intensity to provide the correct exposure. Should you decide on a *shallow depth of field* (e.g. to isolate subjects) lower light levels (intensities) are normally needed. If you need *deep-focus shots,* in which much of the scene is sharply focused, a considerable amount of light will be necessary; this may involve power-consumption and heat problems.

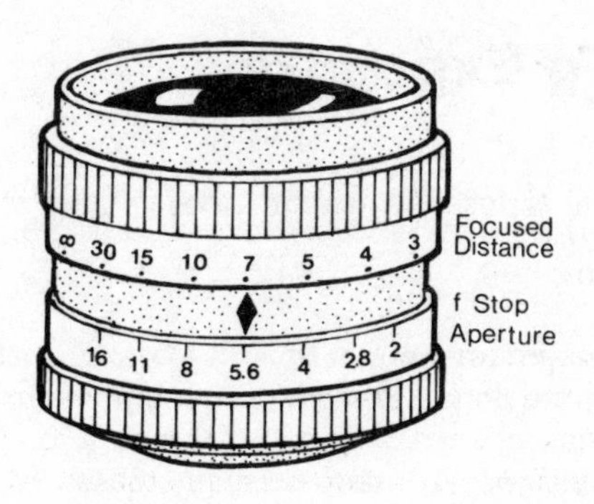

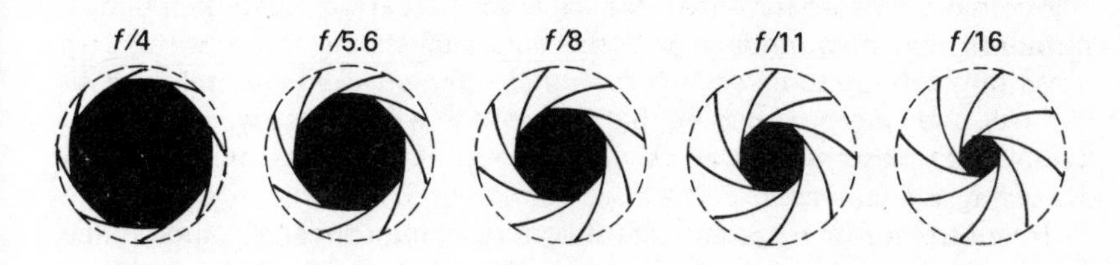

Adjustment
The lens aperture is adjusted by turning a ring on the lens barrel. The *f-stop* number opposite the marker shows the aperture selected. As the lens is stopped (irised) down, its aperture becomes smaller. (Light-transmission is reduced and depth of field increases.)

F-stops and transmission numbers
Lens apertures are often calibrated in a standard series of calculated markings such as: *f*/1.4, 2, 2.8, 4, 5.6, 8, 11, 16, 22. Transmission numbers indicate the amount of light passing through a lens system at various apertures, and can be taken as equivalent to corresponding *f*-stops.

Light transmission
The effect on the amount of light passing through the lens system, on changing to a new aperture, equals:

$$\frac{(\text{old } f\text{-number})^2}{(\text{new } f\text{-number})^2}$$

Thus from *f*/4 to *f*/8 the amount of light received by the camera tube is reduced by $4^2/8^2 = 1/4$.

Light change
Opening a full stop increases the lens aperture to admit twice the original light e.g. *f*/8 to *f*/5.6. Opening half a stop increases the lens aperture to admit half as much light again as originally, e.g. *f*/8 to *f*/6.3.

Controlling Exposure

A picture is 'correctly exposed' when the tones you are most interested in are clearly reproduced.

Light levels

A camera needs a minimum amount of light to produce clear-cut pictures with good tonal gradation and with minimum video noise ('snow') speckling the picture.

Under *inadequate lighting* video noise increases excessively, tonal quality deteriorates and spurious effects develop (e.g. lag, shading). All tones appear unduly dark, and the darkest merge together lifelessly. You cannot compensate for under-exposure by increasing video gain (amplification), since this often emphasises such defects.

When a camera tube receives *too much* light, picture tones pale and the lightest ones merge together, 'blocking-off' to a detailless white; shadow details may become clearer, however. You cannot reduce over-exposure by video adjustment.

To control the average amount of light reaching the camera tube, select an appropriate lens aperture. Remember that this also alters depth of field.

Under high-intensity illumination, you may have to introduce a *neutral-density filter* (page 106). Indoors, you would normally reduce light levels to suit the situation.

Tonal contrast

Varying surface tones and textures, light and shade, often creates very strong contrasts in everyday scenes. Although one's eyes can clearly discern tones throughout a 100:1 range (the lightest 100 times as bright as the darkest), the camera is restricted to a range of around 20:1. Beyond these limits, lighter tones merge to white and darker tones to black.

The simplest way to control picture tones is to keep tonal extremes out of shot. Sometimes you can remove them (e.g. replace a white table cover) or modify lighting (illuminate shadow; shade-off light surfaces).

Exposure

A video picture is normally exposed by looking at a TV monitor (or viewfinder) and adjusting the lens aperture until subject tones are reproduced well. Exposure is usually adjusted for faces, but if you want modelling in extreme tones also you have to compromise.

No automatic 'exposure-indicator' or auto-iris adjustment knows which tones are *important to you.* It adjusts exposure (*f*-stop) to suit averaged-out tonal values. If bright areas come into shot, it reduces exposure but may degrade main subject tones. Under difficult conditions, such exposure aids can be invaluable, to prevent bad over- or under-exposure caused, for instance, by moving from a bright exterior to dim interior. Otherwise you would have to compensate manually by altering the *f*-stop – in this example, by opening up to prevent an under-exposed interior.

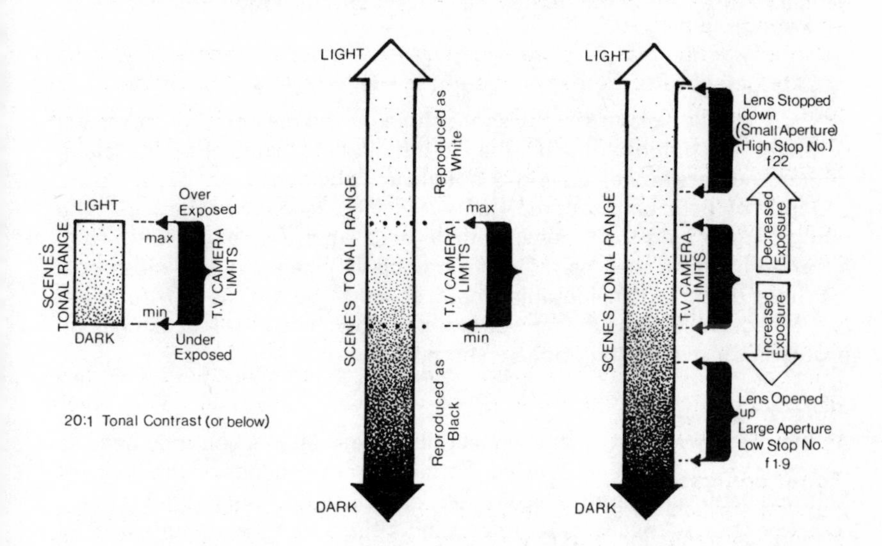

Controlling the exposure
The camera tube can reproduce only a relatively limited tonal range accurately. If scenic tones exceed the camera tube's limits, detail and tonal gradation in the lightest and darkest tones are crushed out.
As you adjust the lens aperture (*f*-stop), the narrow tonal range accepted by the camera moves up/down the scenic tonal scale, selecting which part is to be reproduced successfully.
Opening up the aperture progressively improves shadow detail, but increasingly over-exposes light tones.
Stopping down progressively improves clarity in lightest tones, but darker tones merge.

Careful focusing directs your audience's attention.

Focusing Principles

Strictly speaking, a lens only gives an absolutely sharp image for subjects at the distance to which it is focused (the focused plane). In practice, however, things nearer or further away still look reasonably sharp. This sharpness zone, known as the *depth of field,* varies with the focused distance and the lens's focal length (hence the lens angle), also the *f*-stop (lens aperture). Alter any of these and this sharpness zone gets deeper or shallower.

Depth of field

In a very detailed picture (e.g. a shot of foliage) you may be very conscious of sharpness fall-off outside the 'depth of field' zone. But for plain, unpatterned areas, defocusing is not always obvious.

Depth of field can become quite restricted (shallow) when using a narrow-angle/long-focus lens, with a large aperture (e.g. *f*/1.9). Then accurate focusing may be difficult, particularly when shooting close-ups.

Where there is considerable depth of field, though, as with a wide-angle/short-focus lens that is well stopped-down (e.g. *f*/16) focusing is no problem, for everything appears sharp.

Following focus

As the camera moves, or the subject distance changes, you may need to re-adjust focus. Whether focusing is critical or not depends on the depth of field available and the detail visible. In a close-up, even slight subject movement may necessitate refocusing. But a long shot can usually include considerable movement, without its passing outside the focused zone.

Unless you are using a specially designed 'macro' lens system, you will be unable to focus sharply on anything closer than a meter or so away – this is the *minimum focusing distance (MFD).* For very narrow-angle/long-focus lenses (e.g. 5°) the MFD may be considerably more.

Where to focus

Although you may occasionally defocus a shot deliberately (e.g. to suggest dizziness), you normally want to keep the main subject sharply focused. Anyone can tell whether a printed page is accurately focused, but some subjects are less well defined and therefore much more difficult to focus. In this type of situation you must avoid accidentally focusing on something nearby that you can sharpen on (e.g. detailed background) while the main subject itself remains 'soft'!

When shooting faces close-up, use the eyes (sometimes teeth or hair) to judge focus accuracy. If necessary, rock the focus control to and fro to get the sharpest image. In longer shots, costume detail is usually easier to discern for a focus check.

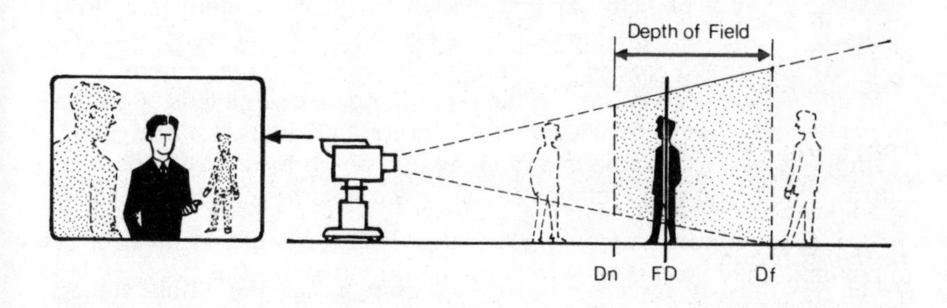

Types of focus control
1. Spoked capstan knob. 2. Twist grip. 3. Camera focus handle.

Depth of field
Within the *depth of field* a lens subjects appear sharp, although maximum sharpness occurs at the focused distance (focused plane), FD. Outside this zone (nearer than D or further away than D) sharpness falls off rapidly and subjects become defocused.

Focusing Problems

Depth of field becomes progressively shallower as the camera moves closer, or zooms in. And here we have the underlying reason for most focusing problems. As we shall see later (pages 58, 100, 102), focused depth may be so restricted that, for example, a pianist's fingers are in focus but not the keyboard; part of a screen-filling shot of an insect may be seen clearly, while the rest of it is a blur.

Selective focus

Ideally, you should be able to select the *f*-stop for the depth of field you need. But in the real world there is often insufficient light for you to stop down enough. Paradoxically, in bright sunshine there may be much more light around than you want and you may have to use ND filters (page 106) to avoid over-exposure. However, where you do have the opportunity to choose the *f*-stop, selective focusing becomes possible.

Deep-focus techniques simply involve stopping down to obtain maximum depth of field, so that everything in the picture is sharply focused. This is ideal when shooting widespread action, or where several subjects are spread around at different distances from the camera.

Shallow-focus techniques, on the other hand, use large lens apertures to deliberately restrict focused depth. Consequently you are able to shoot a single flower, for example, and show it in sharp focus, while blurring surroundings that might otherwise confuse or distract your audience.

Shifting focus

When you have insufficient depth in a shot to see the whole subject clearly, yet cannot stop down, there are several compromise solutions.

1. You can *pull-focus* from one distance to another, e.g. shift focus between persons in a group (but this differential focusing is usually over-dramatic).
2. Alternatively, you may try to *split focus,* adjusting for the best overall compromise – even if this means that nothing is really sharp.
3. Finally, you can take a longer (wider) shot to increase the depth of field.

Hyperfocal distance

If you adjust a len's focus to the *hyperfocal distance* (see opposite page), everything in the shot will be sharp from half that figure to the furthest distance (*infinity*). This *fixed focus* approach helps where focus-following is impracticable, or when you need maximum depth.

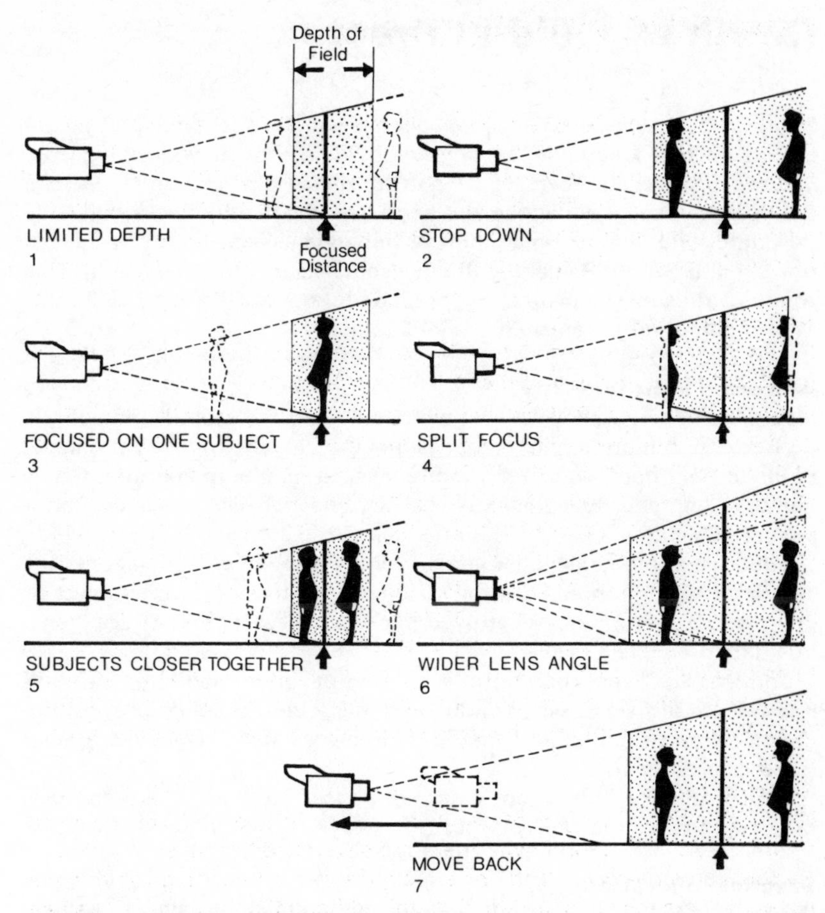

Hyperfocal distance
If a lens is adjusted to its *hyperfocal distance*, everything from half this distance to infinity is reasonably sharp. To find the hyperfocal distance (H):

$$H_{ft} = \frac{(\text{focal length of lens in inches})^2}{\text{lens stop no.} \times 0.002}$$

$$H_m = \frac{(\text{focal length in cm})^2 \times 100}{\text{lens stop no.} \times 0.05}$$

Limited depth of field – solving the problem

1. Depth of field may be too restricted to focus more than one subject properly.
2. *Stop down* – focused depth increases but exposure is reduced.
3. *Focus on one subject* – allow others to become unsharp.
4. *Split focus* – depth of focus is spread between both subjects, making them equally unsharp.
5. *Move subjects* – place both subjects at roughly the same distance from the camera.
6. *Use wider lens angle (zoom out)* – depth of focus increases but subjects now look smaller.
7. *Pull camera further away* – depth of focus increases but shot is smaller.

Depth of Field in Practice

Depth-of-field variations can creep in while you are shooting and concentrating on the action. So be prepared! Remember, if you change the *focused distance* (as camera or subject distance alter), or *zoom* to a different lens angle or choose another *f-stop,* the depth of field will vary!

Roughly speaking, you can think of this zone as extending one-third in front of and two-thirds behind the focused distance (*focused plane*). This zone is correspondingly greater for small-format camera tubes (17 mm/ ⅔ in) than for larger ones (30 mm/1.25 in).

Adjusting light intensity

Most lens systems perform best when only partially stopped down (e.g. *f*/5.6 to *f*/8). Although one might assume that the answer to all problems of insufficient depth is to 'fully stop-down' to minimum aperture, this is not only impracticable under typical lighting but can produce inferior picture quality.

Although some substantial items such as machinery, statues or coins can be lit to high intensities for maximum depth, any living subject or delicate material may be destroyed under such dazzling, hot conditions. Ventilation, too, is a problem.

Conversely, if you open up a lens fully (maximum aperture) for minimum depth of field or when shooting under low-light conditions, picture definition and tonal clarity may deteriorate as various defects appear (flare, aberrations).

Neutral-density filters, too, when introduced to avoid over-exposure, may degrade picture clarity a little.

Practical operation

Depth of field influences all aspects of camerawork. When using a *long-focus lens* (long focal length; narrow lens angle), you soon become aware of how shallow this sharpness zone can be.

Using a *short-focus* lens (short focal length; wide angle) makes focusing easy. In fact, thanks to the greater depth of field, it may hardly need adjustment. But as we shall see, the distortions it can introduce are considerable (page 42).

As a zoom lens has a variable focal length (adjustable lens angle), depth of field changes as you zoom or select another angle. As you zoom in, depth of field lessens; as you zoom out, it increases. So if you shoot someone walking up to the camera, for instance, the zoom setting determines how critical focusing is and how much you have to readjust the focus control to keep up with the movement.

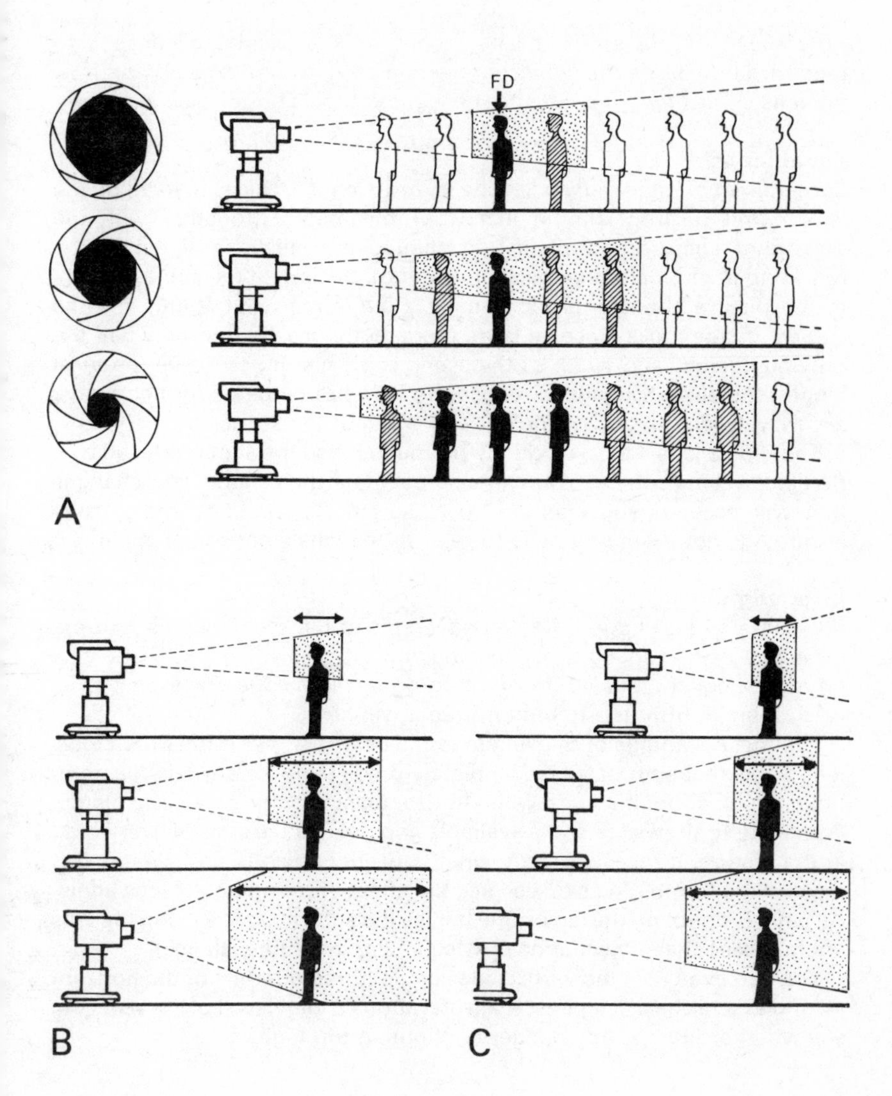

When depth of field changes
A. *Lens aperture.* As the lens is stopped down, depth of field increases.
B. *Lens angle.* As the lens angle increases (shorter focal length), depth of field increases.
C. *Camera distance.* The further away the camera focuses, the greater the depth of field.

Focal Length and Lens Angle

How much of the scene a lens covers at a particular distance, and therefore how large the subject appears on the screen, depends on how the lens's *focal length* relates to the camera tube's target size.

Focal length

Cameramen often identify a lens by its *focal length.* This is marked on the lens barrel or front rim in inches or millimeters (together with the maximum aperture/*f*-stop). If you change the lens's focal length (by replacing it, or by altering the zoom setting), the size of the subject image in the picture changes proportionally. *Double* the focal length and the subject image looks twice as large (apparently nearer), but only half the previous height and width of the scene is visible. If you *halve* the focal length, the subject image is reduced to half size (appears further away), showing twice the original height and width of the scene.

Knowing a lens's focal length helps you to judge the shot produced by a particular camera tube. You can also estimate the relative size changes that will occur if you alter the focal length. To estimate shots really accurately, however, you need to refer to the lens's horizontal angle.

Lens angle

The camera lens shows a four-sided wedge of the scene; with a four-by-three rectangular shape or aspect ratio (indicated as 4:3 or 1.33:1). So if its *horizontal* coverage is 40° from left to right frame-edge, for example, its *vertical* angle from top to bottom frame will be 30°.

The real advantage of thinking in terms of *lens angles,* rather than focal lengths, is that shot detail is completely predictable for any lens/camera-tube combination. If you draw the horizontal lens angle on a scale plan, it immediately shows the shots available and the relative sizes of everything in the picture. It reveals exactly what appears in picture and what is lost beyond the frame. You can also see the effect of changing the lens angle. As before, doubling the lens angle halves the subject size. Halving the lens angle makes the subject appear twice as big (half the distance).

You can even take the vertical lens angle (three-quarters of the horizontal angle) and check it against scale *elevations* (side-views) of the scene, to see whether there is any danger of shooting too high.

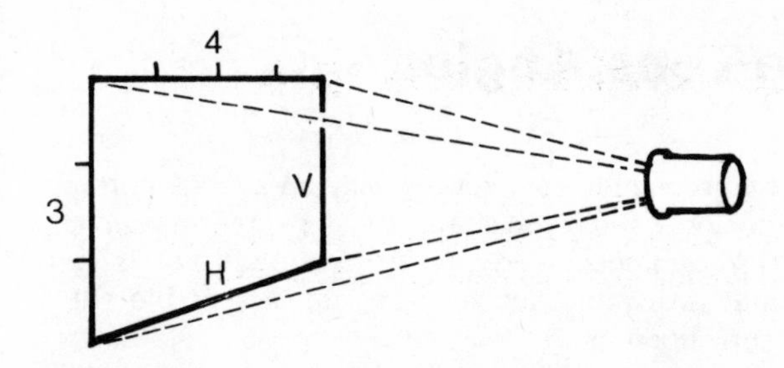

Angle of view
The video camera's lens sees in 4:3 proportions. So its vertical angle of view is three-quarters of its horizontal angle. The longer the focal length, the narrower the lens angle.

Camera tube sizes compared
Camera tube size: 18mm (⅔in)
Image format: 8.8×6.6mm (0.35×0.26in)
'Normal' lens focal length about 11mm (0.4in)

Focal length		*Lens angle*	
mm	in	Horizontal	Vertical
6.0	0.24	72.5°	57.5°
9.0	0.35	52.0°	40.0°
10.0	0.39	47.5°	35.5°
12.5	0.5	38.5°	29.5°
25.0	1.0	19.0°	14.5°
55.0	2.0	9.0°	7.0°
110.0	4.3	4.5°	3.5°
235.0	9.3	2.0°	1.5°
550.0	17.7	1.0°	0.75°

Camera tube size: 25mm (1in)
Image format: 12.8×9.6mm (12.8×9.6in)
'Normal' lens focal length about 16–20mm (0.6–0.8in)

Focal length		*Lens angle*	
mm	in	Horizontal	Vertical
12.5	0.5	54.0°	40.0°
25.0	1.0	27.0°	20.0°
50.0	2.0	13.5°	10.0°
75.0	3.0	9.0°	7.0°
150.0	6.0	4.5°	3.5°

Camera tube size: 30mm (1.25in)
Image format: 17.1×12.8mm (0.67×0.5in)
'Normal' lens focal length about 20–25mm (0.8–1in)

Focal length		*Lens angle*	
mm	in	Horizontal	Vertical
25.0	1.0	36.0°	27.0°
35.0	1.5	25.0°	19.0°
50.0	2.0	18.0°	13.0°
90.0	3.5	10.0°	7.5°
135.0	5.5	7.0°	5.3°
225.0	8.5	4.0°	3.0°

A choice of angles for more flexible shooting.

Varying the Lens Angle

You can shoot an entire production very successfully on a fixed, normal-angle lens (e.g. 25°). For closer shots you simply move nearer (the subject appears larger and surroundings are gradually excluded). For longer shots, you move further away (the subject takes up less of the shot, showing more of the surroundings).

If you are recording intermittently from one viewpoint at a time, this technique poses few problems. If you are shooting continuously, however, frequent camera moves to vary shot size can look very fidgety. When intercutting between several cameras, there is also the danger that one camera will come into another's shot.

A range of angles available

By changing a lens's focal length, you alter how much of the scene fills the screen from your camera position. Shoot a street from a rooftop on a 50° lens (short focal length) and you get a wide overall view. Use a 5° lens (long focal length) instead, and only one-tenth of the original shot height/width now fills the screen. You may take a close-up of a poster that was previously scarcely discernible. These particular angles (5°, 50°) are typical limits for a 10:1 zoom lens.

Fixed (*prime*) lenses are made with angles from as little as ½° to 70° or more; their coverages are usually chosen to provide a succession of 'standard' shot sizes (page 48). Wider lens angles are available for special effects, including 'fisheye' lenses of 140°–360° (all-round view).

No substitute for camera moves

It might seem rational, if you want to alter shot size, to simply keep the camera still and change the lens angle (focal length). This avoids all the inconvenience of moving the camera. On a zoom lens you can do this simply by moving a lever. Some people do just that! But as you will see, this technique can create very undesirable side-effects: perspective distortions and camera handling problems. Moving a camera using a normal lens angle can be much more effective artistically.

There are, however, several good practical reasons why the lens angle needs to vary when shooting most productions (page 40). In fact, without the adjustable lens angles now available, many of today's camera techniques would not be possible.

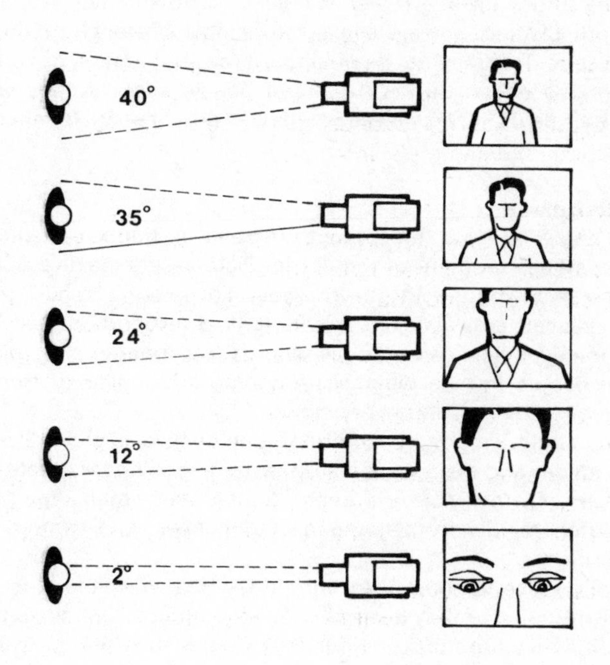

Coverage changes
How much of the scene is covered by the lens (its angle of view) depends on its focal length, relative to the camera tube's image size. As the lens's focal length is changed, its angle of view alters correspondingly. For example, if you use a lens of one-third the focal length, the angle of view becomes three times as wide. Three times as much of the scene is covered but subjects appear a third of their previous size in the shot.

Lens Angle and Perspective

If you are seated so that you can just see the finest detail on a TV screen, your eyes are at roughly 25° to the picture edges. If the scene has been shot by a camera with a corresponding lens angle (say 20°–27°), perspective appears completely natural.

When the camera uses a much wider or narrower lens angle, the apparent depth, distance and scale in a picture look different from those of the original scene. This may not be obvious, it may not matter, or you may actually want the effect. But as a general policy, shoot scenes with a 'normal' lens angle (20°–27°), unless you have a good reason for choosing another.

Forms of distortion

As the lens angle *narrows* (focal length increases), the screen shows a smaller and smaller segment of the scene. Pictorial perspective appears flattened, distance and depth compressed. Foreground-to-background distance is reduced and distant subjects look disproportionately large. This 'telescope-eye' view can produce strange 'cardboarding' if you take close-ups of people from a distance. Anything moving to or from the camera seems to cover ground very slowly.

As the lens angle *widens* from normal, perspective appears exaggerated. Space, depth and distance are emphasized, looking far greater than they really are. Distant subjects seem unnaturally small. Yet people appear to stride rapidly from far to near distances, and camera dolly speeds seem faster.

Theoretically these various distortions arise whenever the lens angle varies from 'normal', but they become more pronounced below about 10° and above 30°. They are most obvious in scenes with strong perspective (e.g. architecture) and least apparent where there are few visual clues (e.g. open spaces).

Deliberate distortion

You can use a narrow-angle lens to create a compressed, crowded-in effect; to group together distant and nearby subjects (pulling together a straggling parade, perhaps, that stretches away into the distance); to reduce the impression of depth and space in a shot.

A wide-angle shot can make even cramped confined spaces seem much bigger on the screen. Quite limited locations can look unnaturally spacious (e.g. elevators) and you can also make modestly sized settings look luxurious. Wide-angle lenses can also be used for dramatic effects, e.g. low-angle shots to emphasize gestures to camera.

Viewing distance
The distance from which you watch a TV picture should vary with the size of the screen. Too close and no extra subject detail is visible. Too far from it and you cannot see all available detail. To see maximum detail, a distance of 4–6 times picture height is often recommended. The screen is at an angle of about 20°–27° to the eye. When the camera lens's horizontal angle is roughly similar, perspective in the picture appears natural.

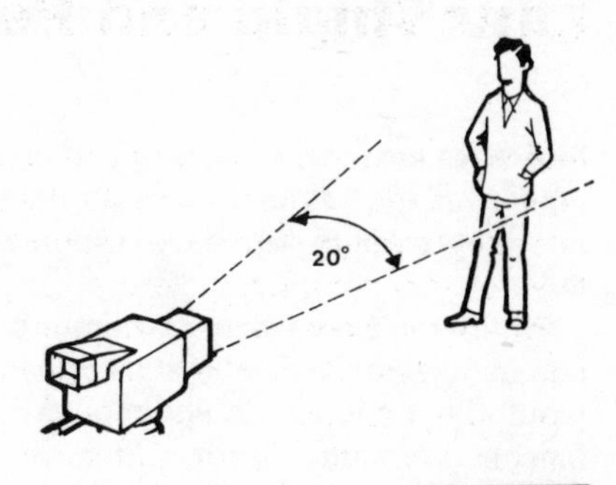

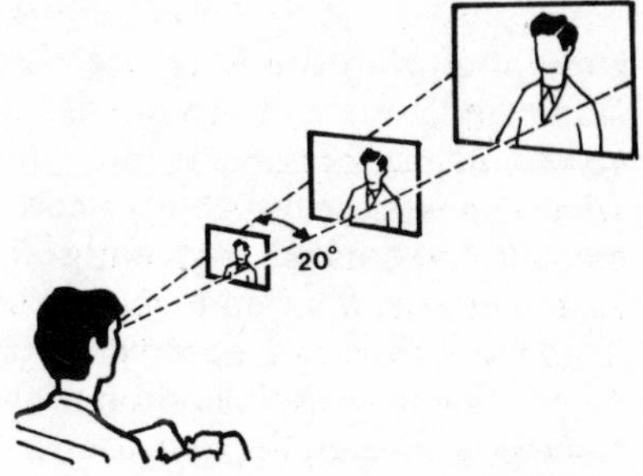

Natural perspective
If you watch the screen from too far away, or a wide-angle lens is used to shoot the scene, depth and distance appear exaggerated in the picture. Viewed too closely, or shot with a narrow lens angle, depth and distance are compressed and distant subjects look unnaturally large.

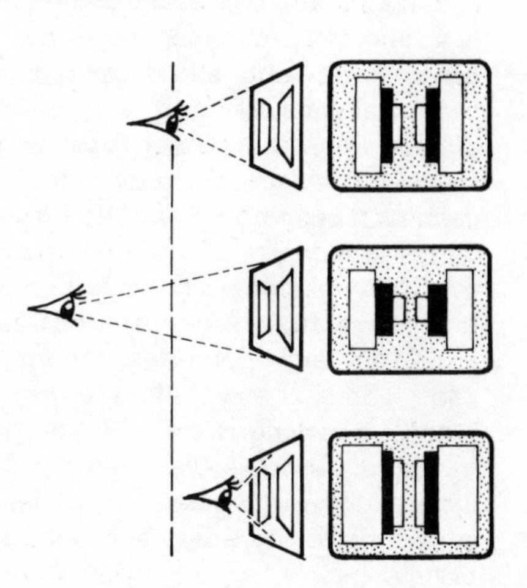

Why Change The Lens Angle?

Although the impressions of distance, scale and proportion are distorted whenever the camera's lens angle differs from normal, most video (and film) cameras usually have a wide range of angles (focal lengths) available.

This is because under practical conditions carefully chosen lens angles can considerably extend shot opportunities. Consider how limited shots would be if you had to shoot a wide-ranging field event or a view from a balcony with just a normal lens.

How changing the lens angle can help

Non-standard lenses can aid effective camerawork in various ways:

1. To adjust framing without moving the camera. They can slightly widen or narrow the shot to correct tight or loose framing; to include/exclude foreground items and so improve composition; to adjust subject size exactly in the frame.
2. Where shots are otherwise impracticable. Narrow-angle (long-focus) lenses can provide close shots of distant or inaccessible subjects when the camera is isolated (e.g. on a rooftop); when obstructions, such as uneven ground etc. prevent camera movement; when the subject is inaccessible (e.g. behind bars); when a tripod is used. Wide-angle (short-focus) lenses can provide wide shots when space is limited, or whenever a camera cannot move sufficiently far from the subject.
3. To alter the apparent subject-distance/shot-size quickly, where there is insufficient time to move the camera. During fast intercutting, rapid camera repositioning may be impracticable or may involve complicated dolly moves.
4. To avoid camera movement where this would distract performers or prevent an audience (or other cameras) from seeing what is going on.
5. To simplify operations. Zooming may produce smoother, faster, more controllable changes than camera movement. When shooting a flat subject, zooming is simpler and more accurate than dollying to/from it (focus-following problems). Narrow-angle lenses taking close-ups are less likely to come into the picture of nearby cameras showing wider shots. For certain situations, altering the lens angle may improve or reduce the depth of field.
6. To adjust the apparent perspective, or proportions. Using a wider lens angle exaggerates space, and a narrower angle compresses depth. By increasing the lens angle and reducing the camera distance (or vice versa), you can change foreground/background proportions (page 64).

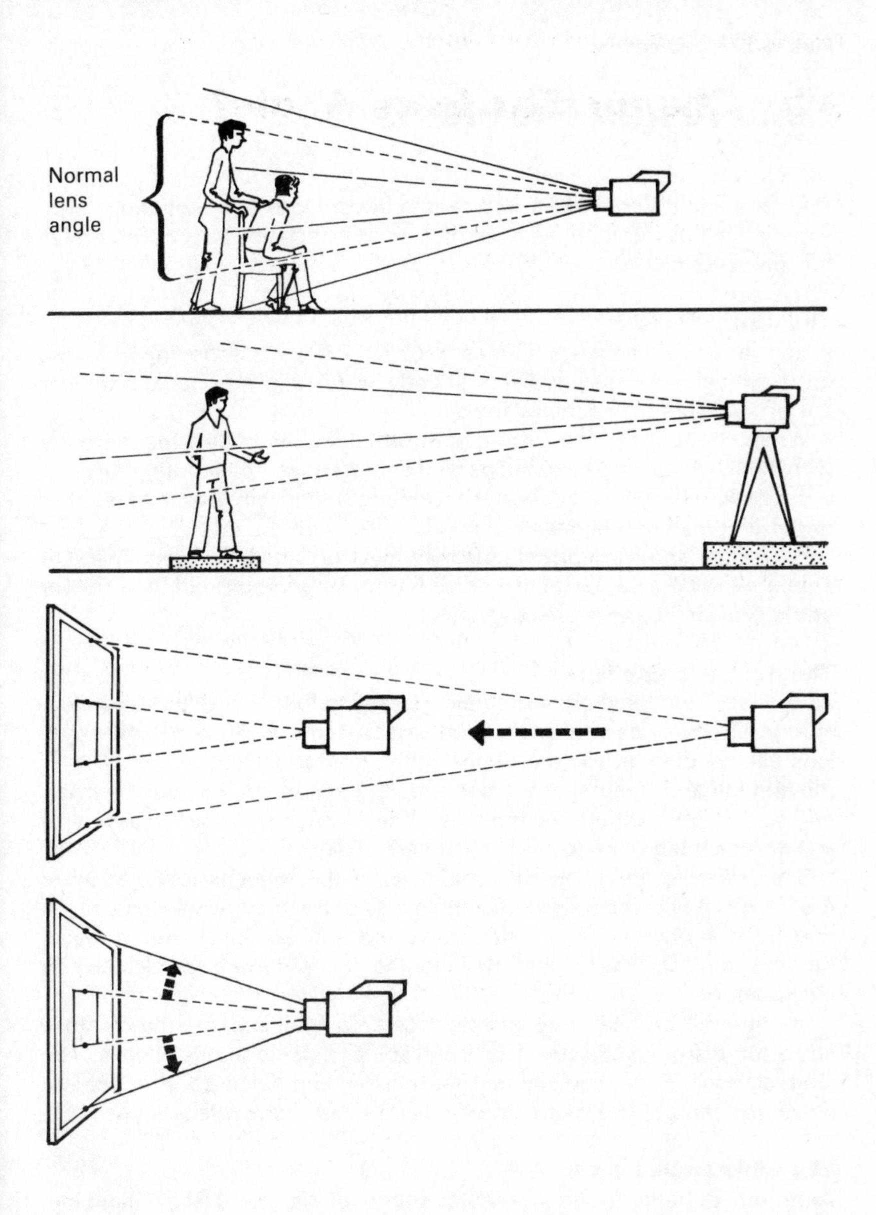

Typical problems

A. When there is insufficient time to dolly, or if movement would distract the performers or hide them from a nearby audience, widen/narrow the lens angle instead.

B. If the camera is fixed or uneven ground prevents it moving, change the lens angle.

C. When a camera dollies in to detail, there can be focusing, framing and movement problems. If the camera zooms instead, these difficulties do not occur.

Lens Angle Problems

Any lens angle you choose has certain advantages and limitations. The experienced cameraman accepts these and selects the best compromise for the occasion.

The normal lens

A normal-angle lens gives few real problems. If you are moving around a lot, however, you may prefer a slightly wider angle to make handling easier and to reduce camera shake.

When shooting in small rooms, a normal lens may give insufficiently wide shots, even when you are backed against a wall, or shooting through a doorway or window. A shorter-focal-length lens will give a wider shot under cramped conditions.

For longer shots, a normal lens may give too small a subject image (it appears too far away) with too much intervening foreground. A narrower angle provides a more effective shot.

The narrow-angle lens

If you have ever used powerful field-glasses to follow bird flight, you will recognize typical narrow-angle problems. A camera with a narrow-angle lens can be difficult to control smoothly. It is often hard to avoid jerky movement and keep the subject accurately framed. For very narrow angles (½°–5°) a tripod becomes essential. It may even be necessary to lock the panning head to prevent camera shake.

Since depth of field becomes shallower as the angle narrows, accurate focusing can become more difficult. Focus control may be very coarse, so that even slight re-adjustment throws the entire subject out of focus. Moving subjects easily pass beyond the focused zone, particularly in closer shots.

In hot weather, heat haze can produce an overall shimmering on close shots of distant subjects. The only solution is to move closer. The characteristic depth-squashing that narrow-angle lenses produce on close-ups can only be overcome by using a less narrow lens angle.

The wide-angle lens

Although popular for their greater depth of field and easy handling, wide-angle lenses do have drawbacks. Subjects often appear too far away (too small in the frame) and foreground space looks excessive.

In close shots, wide-angle distortion is excessive, altering a subject's appearance and grotesquely overemphasizing any movements towards the camera.

Lens flares, toc, are more likely on long shots with wide-angle lenses. In closer shots, cameras cast shadows on subjects and make lighting difficult.

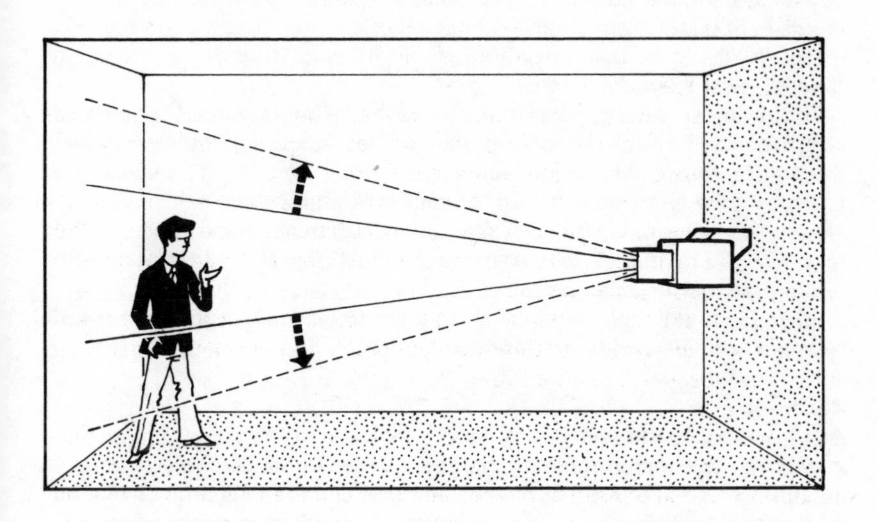

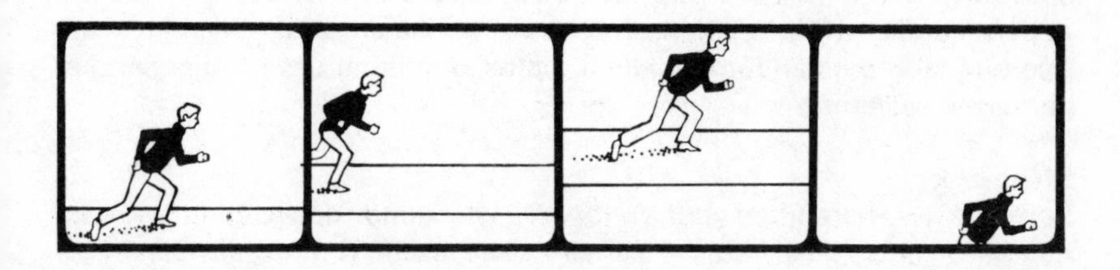

Limited room
If the camera cannot move far enough away to get sufficient coverage on a normal-angle lens, widening the lens angle can help.

Varying framing
When following a moving subject with a narrow-angle (long-focus) lens, you are liable to produce an uneven weaving shot. Constant framing is extremely difficult.

What the Zoom Lens Can Do

Most video cameras today are fitted with a *zoom lens.* This is a complex optical system, providing a continuously variable angle of view as its focal length is adjusted from a narrow angle at one extreme, to a wide angle at the other. It can be held anywhere within its range to perform the same function as a fixed-angle lens.

Zoom lenses vary in design and optical quality. However, many lose performance through simplification, so that focus, light transmission, distortions etc., alter with adjustment. Precision zoom lenses are correspondingly expensive, with greater bulk and weight.

Zoom lenses are made with maximum/minimum angle ratios of 3:1 (e.g. 10°–30°) to 10:1 (5°–50°) or more (e.g. to 42:1). With some designs, a flip-in *extender lens* can be introduced in mid-range (with some overlap) to increase the overall coverage up to three times – but at the expense of reduced definition and light losses.

Zooming

As you increase the lens's focal length (narrowing the angle) the picture *zooms in,* the increasingly magnified image filling the screen with a smaller and smaller area of the scene. Widening the lens angle while on shot zooms the picture *out.* The amount of variation depends, of course, on how much you alter the focal length.

There are several forms of *zoom control,* including a lever on the lens-barrel sleeve ring, a direct pushrod system, and a control mounted on a pan handle (twist grip, thumb-lever, or hand-crank). Each has its operational merits and it is really a matter of getting used to handling a particular system.

Shot box

Some cameras are fitted with a *shot box,* which may be attached to a pan handle or integrated into the camera head itself. This usually has four push-buttons which select pre-set (adjustable) lens angles to suit production needs. So you can change instantly or at a selected speed to a pre-arranged shot. Otherwise you have to adjust the lens angle manually, using a built-in meter or relying on memory while watching the viewfinder.

Where a director has planned lens angles for certain camera positions, the shot box can be adjusted to these chosen angles by using a special graphic (lens chart) or by direct measurement.

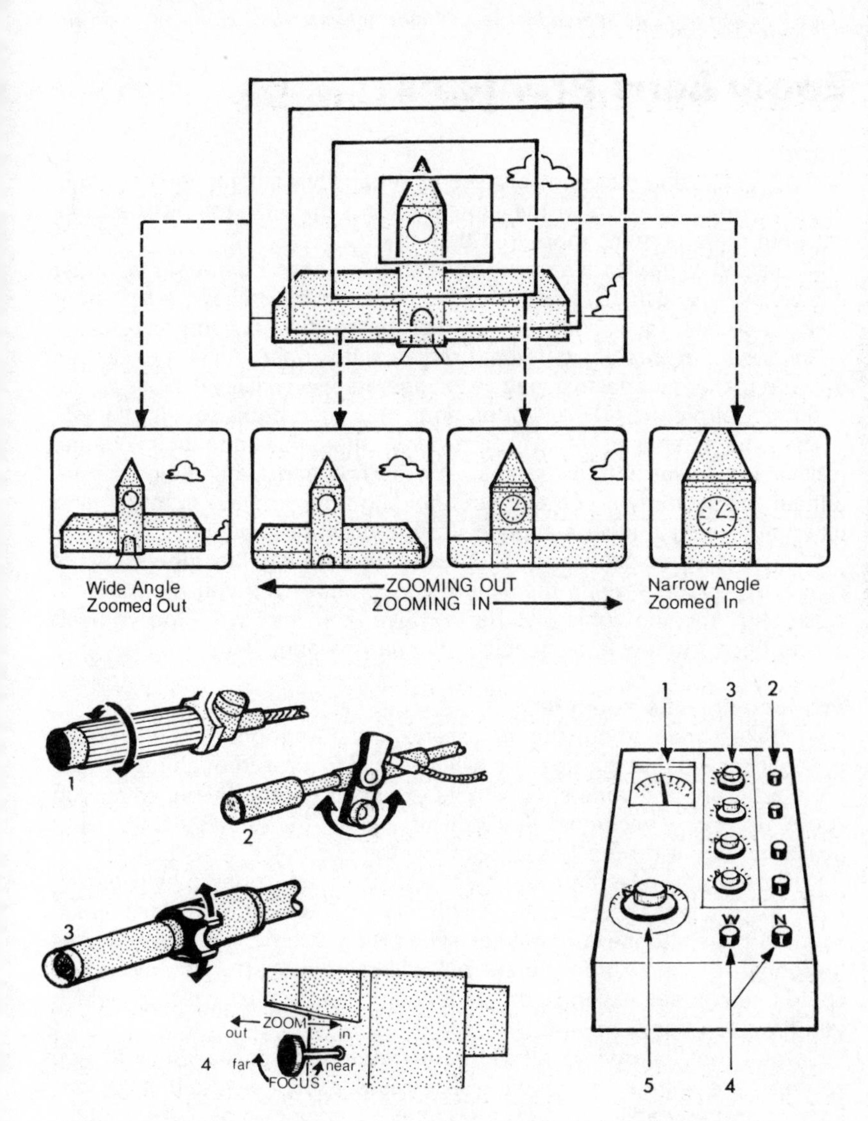

Zooming
Zooming-in progressively fills the screen with a smaller section of the scene (narrowing the lens angle; increasing the focal length).

Forms of zoom control
(1) Twist grip; (2) hand crank; (3) thumb-lever; (4) push-rod (also *focuses*).

Shot-box
Located on a pan-bar or in the camera head, the shot-box pre-selects lens angles.
(1) Meter indicates focal length/lens angle; (2) angle-select buttons; (3) pre-set angle for button selection; (4) move to widest/narrowest angle; (5) zoom speed.

Zoom Lens Problems

At first sight, using the zoom lens seems simplicity itself, for you have only to change its angle to alter the shot, and focus up when the picture looks soft. But there is much more to it than that!

As discussed earlier, a *narrow-angle lens* can have frustrating features: it is sensitive to camera shake; depth of field can be shallow; the point of focus is critical; focus control is coarser; perspective is compressed.

A *wide-angle lens* has different problems. It is often difficult to see the focused plane; perspective may be exaggerated, especially in close shots; geometry at picture edges is often poor; it is susceptible to light flares.

The characteristics of a zoom lens alter throughout its range, particularly towards its extremes. At a narrow angle, handling is more difficult, particularly if you are hand-holding the camera. Zoom out, and handling is easy but wide-angle distortion becomes evident.

If lens angles are continually altered during shooting, the impressions of space, depth and scale in the picture vary arbitrarily. If you zoom in to a close shot and the subject starts to move, you may well find yourself attempting to follow it, unsteadily, on a narrow-angle lens!

Pre-focusing the zoom lens

If you take a zoomed-out shot of a scene, and then zoom right into detail, the chances are that the close-up will be out of focus. Although focusing is quite arbitrary in a wide shot with its extended depth of field, when you zoom in it can becomes very critical, owing to the restricted depth available. Any error is all too obvious in the picture.

Wherever possible, therefore, try to anticipate any zoom-in by sneaking a *pre-focus check* beforehand (i.e. zoom in . . . focus hard on close-up . . . zoom out again to the current wide shot). Then, when the moment comes to zoom, you know that the subject will remain sharp. Otherwise you could find yourself making a dramatic zoom . . . into a fuzzy picture, which you have to refocus on-air.

Some zoom lens designs have poor tracking, i.e. focus varies as you zoom. There is little you can do in such cases, except ensure that the lens itself is properly adjusted and correct any variations as best you can.

The need to pre-focus
In a wide-angle shot, considerable depth of field makes it difficult to see exactly where the lens is focused most sharply.
As the lens angle narrows, depth of field becomes more restricted. So when zooming-in you may find that the focused plane is not at the subject, and it is out of focus in the close-up.

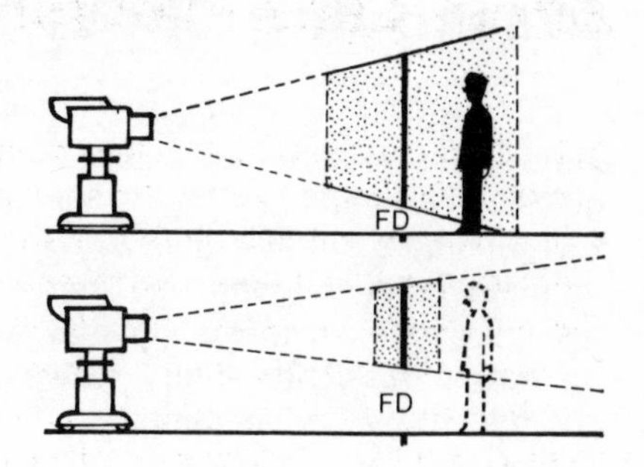

Changing perspective
Zooming-in on a three-dimensional scene causes changes in apparent perspective and proportions. Distance and depth become accentuated at wide angles but flattened and compressed as the lens angle narrows.
Zooming-in on a *flat* surface results only in a change in magnification. Proportions do not alter.

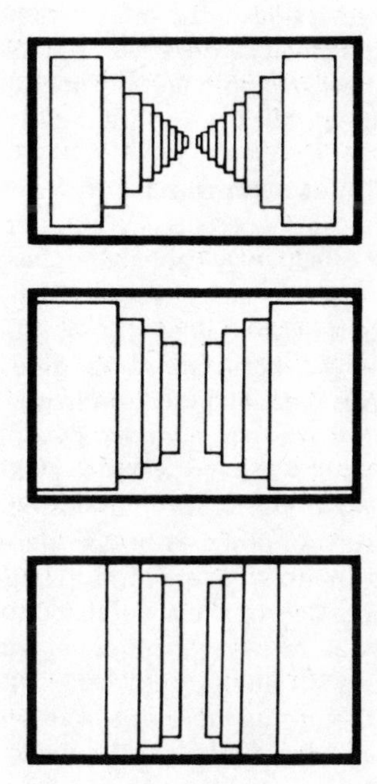

Classifying Shots

The camera does so much more than just 'take a picture' of a situation. It gives your audience a certain impression of a subject and its surroundings. Shoot it one way and the subject can appear important, dominating its environment. Shot from another angle, the subject often becomes quite incidental. It may even be overlooked. So we need a method of classifying shots, to help us organize and arrange how we are going to shoot any situation.

Defining the shot

We usually assess any shot according to how much of the screen the subject fills – how close it appears to be. As far as shot classification is concerned, it does not matter whether you obtain a close-up, for example, by using a close camera with a wide-angle lens or a more distant camera with a narrow-angle lens. The overall *effect* is different (perspective, proportions), but the shot *size* can be similar.

General classifications

Although the TV screen is comparatively small, it can show a wide range of shots effectively, especially closer viewpoints – although it does not present panoramic or spectacular situations particularly impressively. Each type of shot has certain productional purposes.

Very long shots (vista shots) give an impression of space and distance, and help to establish location (e.g. a seashore). It can show where people are within a scene (e.g. on a rock) or show how one action group relates to another (e.g. a hidden watcher awaiting his victim).

In a *long shot (full shot),* the camera shows a distant view of the subject and its surroundings. It helps, therefore, to show where action is taking place, to create an overall mood and to follow fast or wide-ranging action. Long shots do not show detail, so for this you need to use closer shots. A shot may show several people at a table, for instance, sufficiently close to see relevant details (gestures, expressions), yet sufficiently distant to convey the atmosphere of the surroundings.

A *close shot (tight shot)* usually fills the screen with greater detail, pointing out particular features, laying emphasis, dramatizing perhaps. By excluding the surroundings, the shot concentrates attention, perhaps preventing the audience from seeing other things happening nearby and being distracted from the main subject.

Shot size

You can get the same kind of shot from a close position using a wide-angle lens, or from a more distant position with a narrower angle. Depth of field remains identical for the same size shot, although perspective appears different.

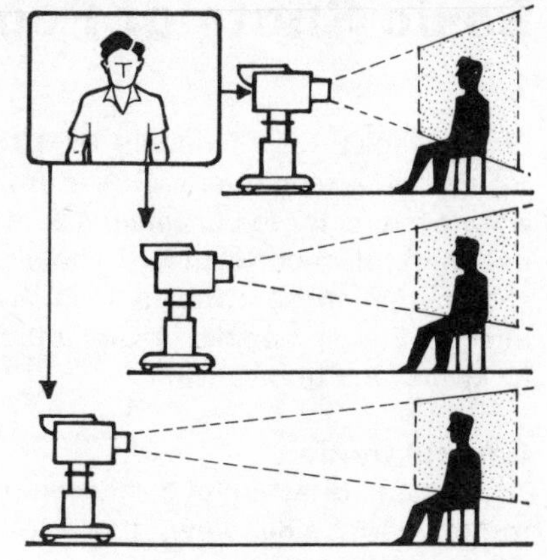

General classifications

Broad terms are often sufficient to indicate the sort of shot required.

(a) A long shot/full shot shows overall action in a distant view.

(b) A wide shot/cover shot shows performers and their surroundings in broad detail.

(c) A very long shot/vista shot reveals general location.

(d) A close shot/tight shot concentrates on detail.

(a)

(b)

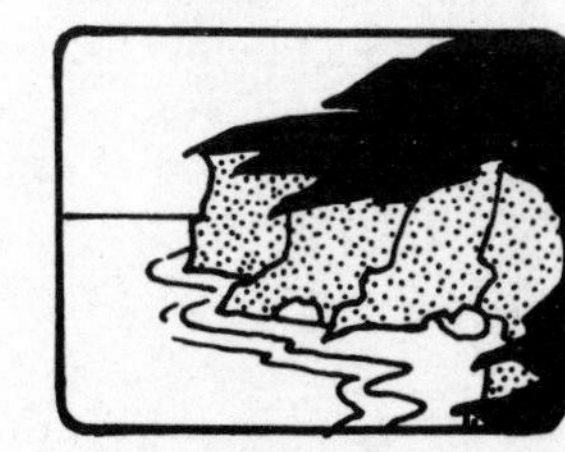

(c)

(d)

Basic Shots of People

Effective pictures of people tend to follow a series of regular, easily recognized arrangements. These provide convenient quick references that enable a director to indicate the shot he wants in just a few words. But do not think of such 'standard shots' as simply routines. Experience has shown that these compositions provide the most artistically pleasing effects. Frame people in any other way and the shot usually looks awkward and unbalanced.

General terms

The general direction of a shot can be indicated by the broad description: frontal shot, side view, three-quarters frontal, or back (rear) view. Additional indications could be low shot, level shot, high shot, to show height. Sometimes a general indication, such as *'over-shoulder' shot* or *'point-of-view shot'* (POV) is sufficient.

For some purposes it is enough simply to indicate how many people are to be in shot, and the general guide *'single shot', 'two shot', 'three shot'* or even *'group shot'* is used.

Classification

To indicate exactly how much of a person is to appear in shot, the shot classifications opposite are used. Several terms have evolved for each but the shots themselves are universal.

To prevent confusion, though, it is best to use those found locally. Where a director talks about an 'MCU' instead of 'chest shot' or 'bust shot', follow that terminology. After all, the purpose of these classifications is to convey information and in most organizations they have become standardized by custom.

You can remember these arrangements by their framing, such as 'cutting just below the waist', 'just below the knees' etc. In no time at all, you will find yourself automatically thinking in these terms, free to concentrate on other aspects of the action.

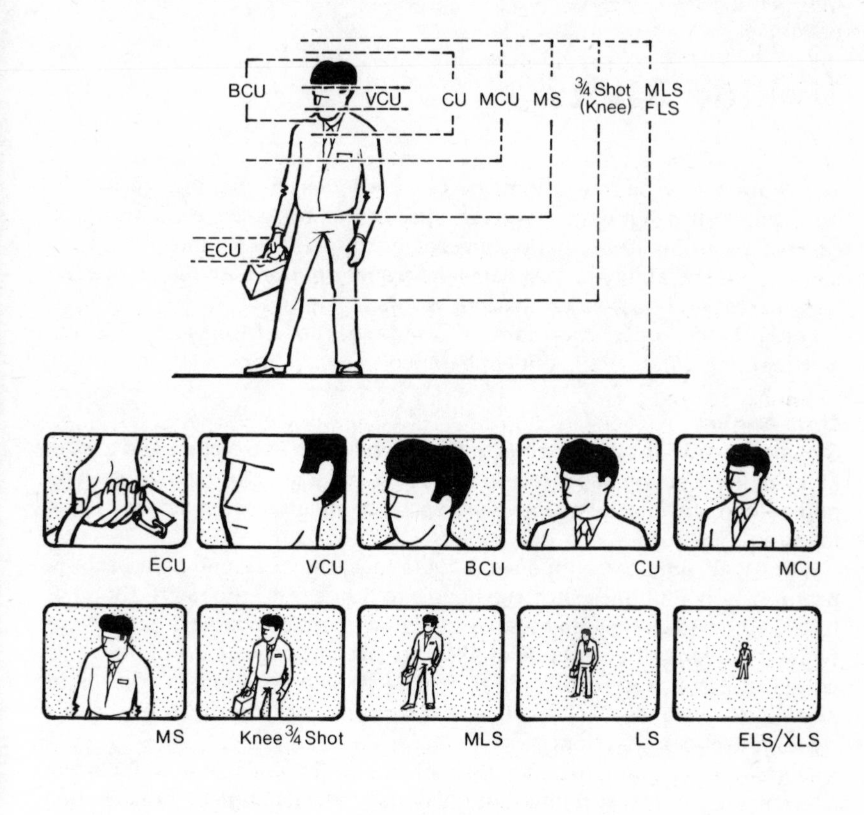

Shots are identified by how much of the subject they include:

ECU Extreme close-up (detail shot) – isolated detail.
VCU Very close-up (face shot) – from mid-forehead to above chin.
BCU Big close-up (tight CU, full head) – full head height nearly fills the screen.
CU Close-up – just above head to upper chest (cuts below necktie knot).
MCU Medium close-up (bust shot, chest shot) – cuts body at lower chest (breast pocket, armpit).
MS Medium shot (mid-shot, close medium shot, CMS, waist shot) – cuts body just below waist.
Knee, ¾ shot Knee shot, three-quarter length shot – cuts just below knees.
MLS Medium long shot (full-length shot, FLS) – entire body plus short distance above/below.
LS Long shot – person occupies three-quarters to one-third screen height.
ELS Extra long shot (extreme LS, XLS).

Only exceptionally do you want to make a person appear misshapen and grotesque.

Unkind Shots

There are a few people whom the camera loves! Somehow, no matter how harsh the lighting, however sparse the make-up or critical the camera's scrutiny, these people always look great! But for most of us, the camera has the ability to misshape and transform our features, until even our nearest and dearest are lost for words.

Let us see how the camera manages to emphasize physical shortcomings and distort the appearance.

Lens angle

Since a narrow lens angle tends to compress depth in a picture and a wide lens angle tends to exaggerate it, it is not surprising to find that these lens angles can do exactly that when reproducing three-dimensional facial features.

If you get too close with a wide-angle lens, the nose and chin become prominent in a full-face shot, the head's roundness is emphasized and the forehead recedes. Results are so grotesque, on a close camera with a very wide-angle lens, that nobody is likely to attempt serious portraiture this way. But if space is restricted, or camera movement limited, you might attempt to use a close wide-angle lens simply to get the shot, despite this cartoon-like exaggeration.

Narrower lens angles than normal are often advocated for correct portraiture. But you will find that really narrow lens angles, for close-ups of distant people, produce unacceptable facial compression. In full-face shots features lose modeling and flatten, the nose appears squashed, the forehead and chin may look more prominent. The effect is of a photo cut-out. Yet it is so convenient when a person is some distance away (particularly if the camera cannot get nearer) to take a close-up with a narrow lens angle and ignore the distortions that must result.

Camera height

The height of a camera's viewpoint can affect a person's appearance in several ways. Elevated or high shots tend to emphasize baldness, plumpness, bosoms, and generally make a person look shorter and less imposing.

Depressed or lower camera viewpoints, on the other hand, tend to emphasize noses, particularly large, uptilted or dilated nostrils. Lower shots also draw attention to scrawny necks and heavy jawlines. A person with a high forehead or receding hair may appear completely bald in the depressed shot.

Background influence
An inappropriate background can draw attention to the subject's less attractive features.

Shooting from above
High-angle (elevated) shots can be unflattering by emphasizing foreheads and baldness, and reducing effective height, which often makes people look weak and lacking in authority.

Shooting from below
Low-angle (depressed) shots can draw attention to other less attractive features – scrawny neck, deep-set eyes, nostrils, and can suggest baldness with a high hairline.

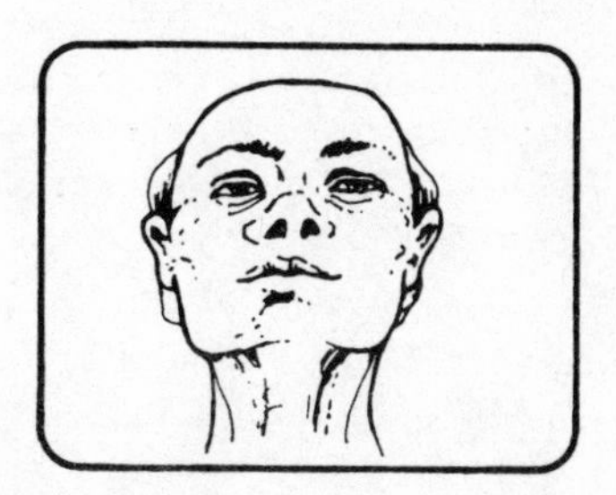

Used to establish environment, broad action, atmospheric effect.

Long Shots

In long shots the location predominates. Normally a long shot is used to show where the action is taking place; to follow wide-ranging or fast action; to show how groups of people or things are interrelated; to reveal an environmental effect (showing how rich, squalid, crowded, empty a place is); to convey information about the location (e.g. sunny day, moonlight through window, candlelight, etc.).

On the small TV screen, long shots give an overall impression; they cannot show detail. Long shots therefore need to be interspersed with closer viewpoints to provide a visually balanced presentation. Whether you use many long shots or introduce them only occasionally depends on the nature of the production. A horse race, for example, continually needs broader views of the action, while in a demonstration of flower arrangement close shots must predominate.

Camera operation

In long shots there is usually considerable depth of field, so focusing poses no problems. However, you are unlikely to be able to pan or tilt the camera much, without shooting beyond the acting area and perhaps including lighting, cameras or other extraneous subjects in the shot.

When you have to use a wide-angle lens to get a long shot (where otherwise there would be insufficient coverage), perspective is exaggerated and straight lines near picture edges are liable to bend or slope unnaturally.

Most camera movements (dollying, trucking) are not particularly effective in long shots, although changes in camera height can still be dramatic.

Lens flare

When you are shooting into lights, or where lamps are just outside the shot pointing towards the camera (backlights), lens flares are a regular hazard. In the color TV picture, these visual blemishes take the form of patches, rays, blotches or veiling over part of the shot. All too obvious on a color monitor, the effect of lens flares may be barely discernible in the cameraman's black-and-white viewfinder.

Space
Where confined space prevents you getting a long enough shot, camera distance can often be increased by shooting in through a door or window.

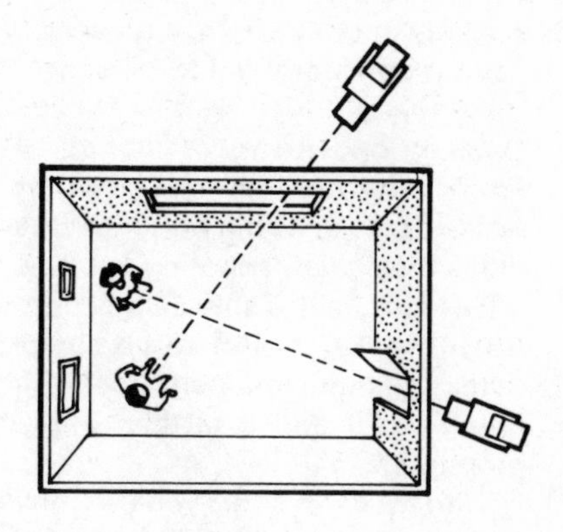

Overshoot
Cameras can easily overshoot past the edges of the action area and include unwanted items in the shot.

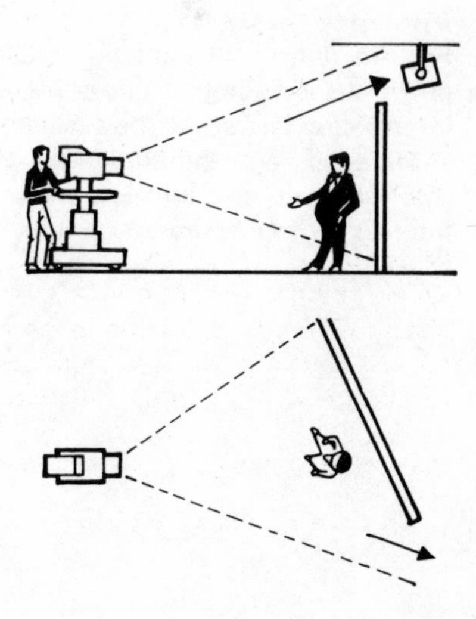

Generally trouble-free, medium shots are probably the most used set-up.

Medium Shots

Quite a range of shots can be called 'medium'. Typically they cover people from full-length to mid shots. The function of a medium shot lies somewhere between the environmental impact of the long shot and the scrutiny of close shots. It presents the performer within his surroundings, so that both combine to influence the audience.

Camera operation

From *full-length* to *three-quarter length* shots you can frame large gestures, such as big arm movements. With this type of shot there are no difficulties when shooting subjects that move around a little.

Depth of field is sufficient at normal working lens apertures (e.g. around *f*/5.6) for the subject to be sharply focused, while leaving background details slightly softened. As the shot tightens, the surroundings become less distinct and lose their impact, so making the subject itself more prominent.

With *medium* shots you can move the camera around smoothly on shot (using a normal to wide lens angle) and introduce effective changes in viewpoint, particularly if your moves are *motivated* by action within the scene – e.g. someone getting up and going over to open a door.

Audience interest

For the director the medium shot provides a safe, uncritical, general-purpose viewpoint. It offers his audience a useful amount of detailed information and so can be sustained for a relatively long time. (While *long* shots encourage the audience's attention to wander around the scene, *close* shots confine the view and present less information, so usually hold interest for much shorter periods.)

Medium shots
These have various important features. They enable the viewer to see a reasonable amount of detail in the subject, while at the same time revealing its surroundings. So the environment, lighting etc. make a strong impact. Medium shots can contain a considerable amount of action without requiring adjustment of the camera's position.

They have a well-deserved reputation for their difficulties!

Close Shots

As you take closer shots of a subject, camerawork becomes more critical. The method you choose to get these shots affects the sort of problems you have.

With a close *wide-angle* lens, the subject may appear misshapen and shadowed.

Use a *narrow-angle* lens – further away, of course – and these problems of distortion disappear. Instead, you may be limited by the lens's *minimum focusing distance (MFD)* and find that when you have moved the camera close enough you cannot turn the focus control any further, to sharpen the image! Coarse camera handling often frustrates subtle framing of close shots, and 'narrow-lens distortion' may flatten modeling.

Depth of field

As depth is so restricted in close shots, you often have to decide whether to accept the situation, split focus, or try to increase depth of field by either of the following methods:

1. By *stopping down* (more light is needed).
2. By taking a *wider shot* (the subject now appears smaller).

Subject movement

The closer the shot, the more difficult it is to shoot movement. Not only can focusing and framing become erratic, but because the subject fills a large part of the screen any inaccuracies are embarrassingly obvious. Although you may sometimes allow parts of the subject to move out of shot, you should normally try to contain all subject movement within the frame.

When taking close shots of a book, pages have to be held quite steady if details are to be seen clearly. For very tight shots (e.g. a watch face), even slight hand shake is disastrous.

Confining the subject

To avoid embarrassment, use a marked position on a table top when possible, where the item can be laid or supported firmly. If details are really tiny (e.g. a postage stamp), it may be better to record a separate *'cutaway shot'* of this (to be edited in), rather than try to catch the shot 'on the fly'.

Shot limits
In very close shots, action should be kept within the frame area, avoiding movements in/out of shot.

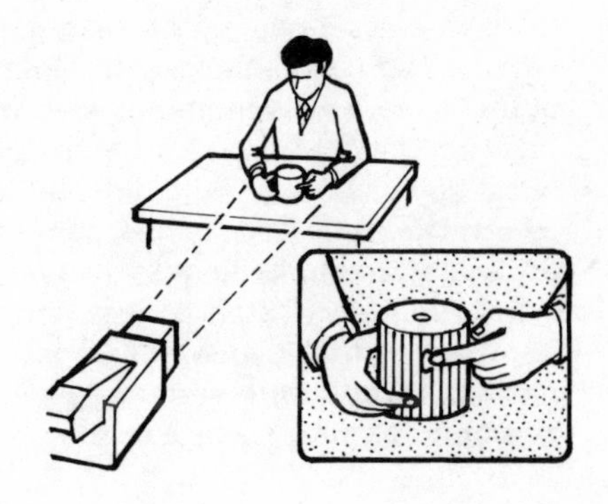

Shallow depth
Although depth of field is limited in close shots, you can make the most of it by ensuring that important surfaces are at right angles to the lens axis. If a flat surface is tilted, parts of it can fall outside the focused zone and become defocused.

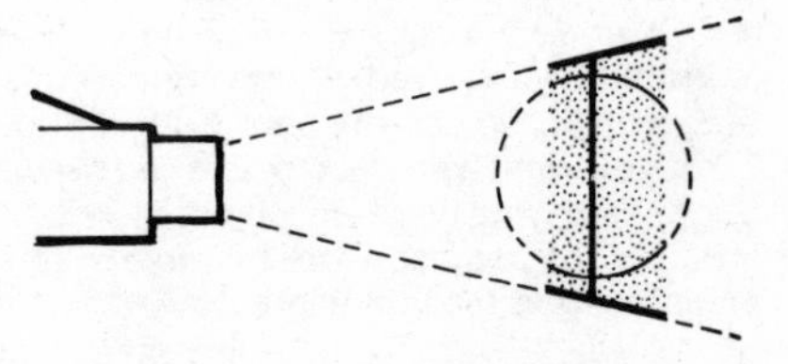

Calculating shots

Knowing the focal length or angle of view of the lens he is using, an experienced cameraman or director can make a shrewd guess at the sort of shot he can obtain. A director who wants shot information during pre-camera rehearsal often uses a portable adjustable viewfinder, showing him the shot-sizes different lens angles will provide. This is fine when there is something to look at. But what do you do when planning and estimating production treatment before a single thing has been built?

Using a scale plan of the setting, the studio or the locale, you can see what will come into the picture, simply by placing a transparent protractor at the camera viewpoint. But what if there is no scale plan?

All you have to do then is to look up the graph and see at a glance the shots you can obtain with different lens angles at various distances. This avoids laborious calculations, pages of tables, or trial and error. To extend the scales, multiply them by two (or more).

Knowing, for instance, that someone is 3 m/10 ft away, it shows immediately that if you want a close-up a lens angle of just over 10° will be needed. If your narrowest lens angle is 20° you have to move the camera nearer, to around 2.5 m/8 ft from the subject, to obtain the shot.

How to use the graph

1. *To see the shot you will get,* draw a vertical line up from the camera distance to the lens angle being used. The shot you get is shown on the left scale.
2. *To see how far away* you need to be from the subject to get a certain shot, choose the shot-size on the vertical scale. Follow a line across to the lens angle you are going to use, then down to the distance below.
3. *To find the lens angle* needed for a certain shot, first draw a line up from the distance scale, then another across from the shot size you want. Where they meet, you have the lens angle needed.
4. *To find the scene width* taken in by the lens, look up from the distance to the lens angle being used. Where it meets this angle, look across to the vertical scale on the left. This shows the shot width. (Shot *height* is three-quarters of the width).
5. If you want a subject to fill a *certain proportion of the screen,* this is easily done. Supposing you want an object to take up one-third of the frame width. Simply multiply the *subject width* by three in this case, and find the camera distance/lens angle needed for that width.

Lens Angle

Horizontal	Vertical
5°	3·75°
10°	7·5°
15°	11·25°
20°	15°
25°	18·75°
30°	22·5°
35°	26·25°
40°	30°
45°	34°
50°	37·5°
55°	40·25°
60°	45°

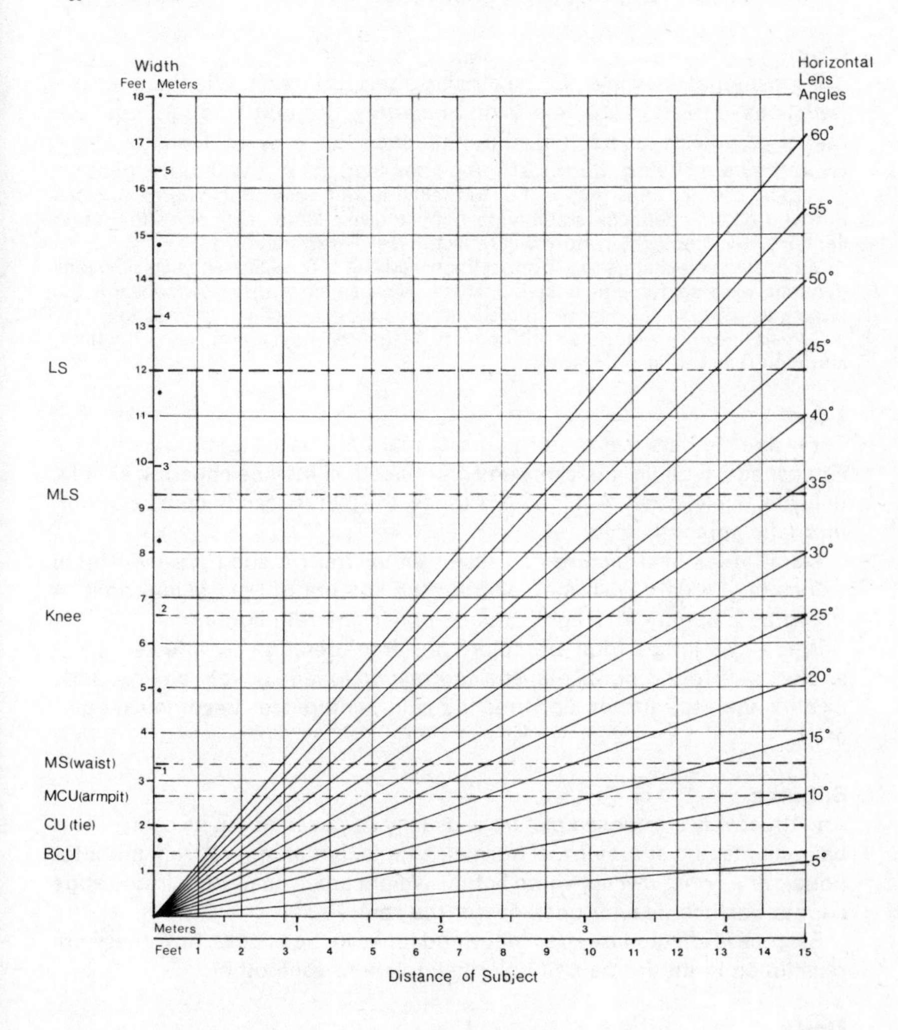

Universal camera set-up graph
The graph shows, at a glance, details of obtainable shots. For greater or shorter distances, just multiply or divide the scale readings.

The Basics of Composition

Although we can only touch here on the underlying principles of pictorial composition, these basics will help you to select and arrange pictures more successfully. Study film and TV programs and you will see them in regular use. With practice, you will soon find yourself composing pictures instinctively, for the most appropriate effect.

Line

Compositional lines in a picture directly affect its impact. Whether they are *real* lines (structural, painted) or *imaginary* (formed through arrangements of objects or people), they influence the viewers' feelings about what they are seeing. Compositional lines lead the eye within the picture.

Vertical straight lines give a shot formality, height, restriction. *Horizontals* can impart breadth, openness, stability, rest. If you take a flat-on shot of a subject that contains such lines, the picture will reflect these characteristics.

Supposing you change to a more oblique viewpoint. The same subjects now look dynamic, exciting, forceful, unstable. That is because these lines now appear in the picture as *diagonals:* lines which are inherently more arresting, interesting.

Curved lines are associated with beauty, elegance, movement, visual rhythm – although they can also be weak.

Tone

Tone directly sets the mood and pictorial balance of a shot. Position a subject against a *light* background and the effect may be cheerful, simple, delicate, lively, open. Against *dark* tones, the picture can become somber, dramatic, forceful, drab.

Tonal areas in a picture can result either from a subject's own tonal values (e.g. a dark costume), or from the amount of light falling on it. A mid-gray box may look light or dark, under different lighting.

When we judge tone or color, our interpretation is affected by a subject's surroundings. Against a *contrasting tone* (e.g. light against dark, or dark against light) or a contrasting color, differences become exaggerated.

Balance

An attractively composed picture is usually *balanced* about its centre. This balance may be *symmetrical* (formal, simple, but comparatively monotonous); or *asymmetrical* (where lighter-weight areas nearer the frame edge counterbalance heavier ones nearer the centre).

Balance is affected by size, shape and tones in the picture, how these are positioned in the frame and how they relate to each other.

Unity

Unity is the principle of arranging subjects within the picture so that they look interrelated or grouped, rather than scattered around as separate items.

Line
Lines creating shapes and patterns direct the eye in the picture. These lines may be *real*, forming an actual part of the scene. Here converging lines draw attention to the distant building. Lines may also be entirely *imaginary* but felt to be there as we look at a picture, as with this 'triangle', which seems to give the subject stability and unity.

Tone
Tones influence how we respond to a picture. Darker tones give an enclosed feeling compared with the open effect of light tones in the second picture. Against darker tones, light-toned clothing is more prominent.

Balance
1. Shots need to be balanced around the picture centre.
2. Lack of balance makes the picture appear attractively unstable.
3. A formally balanced picture with a symmetrical arrangement can look uninteresting and monotonous.
4. By adjusting the size and position of tonal areas in the picture (by careful positioning and framing) a more attractive effect is achieved.

Practical Composition

Although an artist with a blank canvas has a free choice as to how he is going to manipulate shapes, texture, color relationships etc., a cameraman is working in the real world, composing the picture from what is there in the scene.

Sometimes he may be able to rearrange the subject, or reposition features to improve the shot. But a video cameraman's expertise usually lies in seeing suitable visual opportunities, in selecting exactly the right viewpoint and shooting the subject in the most appropriate way. If he does this skilfully, the result will be so convincing that it all seems to have happened naturally. But if a shot is badly composed, it looks contrived, fails to hold the audience's interest, or lets their attention wander.

Adjusting composition

When a director pre-arranges a camera's set-up (and this is what happens in most situations), the cameraman has to compose his shots relative to whatever is available there. At first sight, this might seem pretty limiting, offering little scope for compositional choice. But in practice, visual emphasis can be altered in a number of ways:

1. *By adjusting subject size* – you can make a subject dominate the scene, or recede into the background. Certain foreground features can appear strong, quite incidental, or left out of shot entirely, so that the viewer does not even know they are there.

2. *By adjusting framing* – you decide exactly what is visible to your audience. You can include, exclude or emphasize other subjects in the scene.

To build up tension you might, for example, frame a shot to reveal that someone is walking unsuspectingly towards a hidden escaped animal. Alternatively, you might deliberately exclude the animal, creating a sudden shock as it leaps out onto him. Even slight alterations in framing can change the entire compositional balance of a shot.

3. *By adjusting the lens angle and camera distance* – you can alter the relative proportions between the subject and its surroundings (page 40).

4. *By choosing the lens height carefully* – you can strengthen or weaken a subject; arrange foreground items to frame the shot; or shoot over (or under) foreground obstructions.

5. *By moving sideways* – you may prevent a nearer subject from obscuring another one further away; or you might avoid something intruding into the shot.

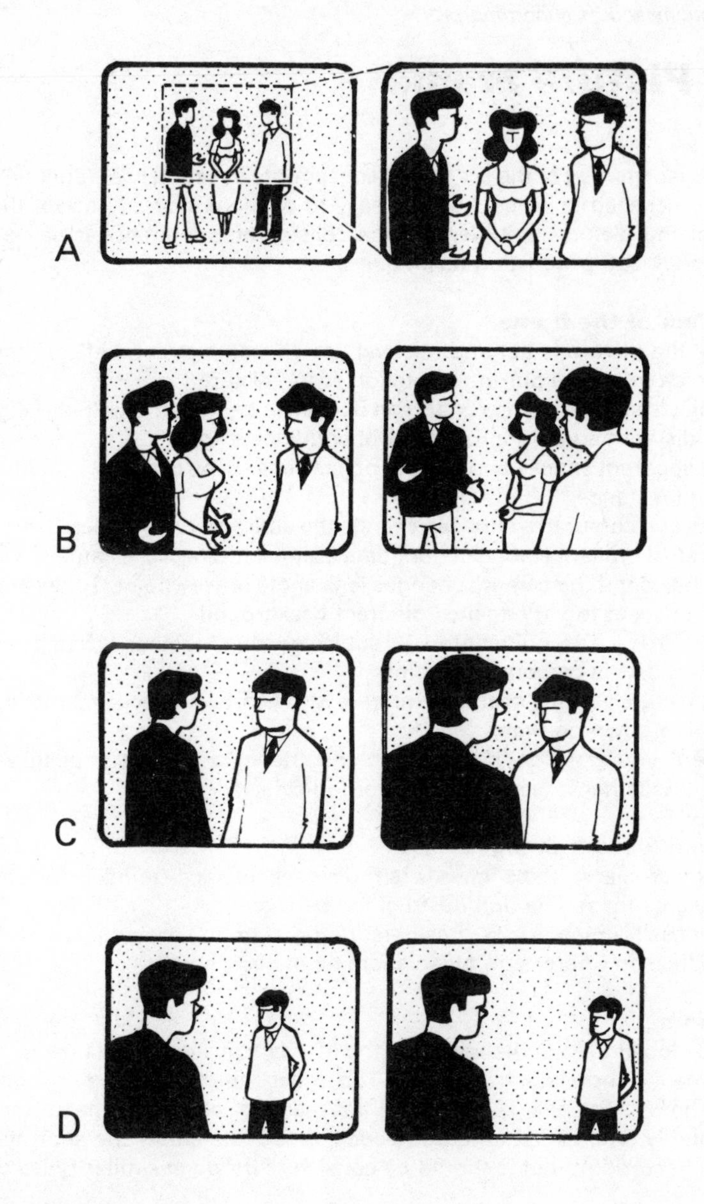

Adjusting composition

A. By adjusting shot size, the relative visual impacts of the subject and its surroundings are altered.
B. By slightly changing the camera position (trucking), the picture balance can be altered.
C. By adjusting the lens angle and camera distance, proportions can be changed.
D. By moving to a new viewpoint, the subject grouping (visual unity) can appear quite different.

The Picture Frame

All shots isolate part of the scene and concentrate the audience's attention on those selected features. Yet it is easy to overlook how important the frame of the picture really is, and the considerable influence it has on camerawork and production techniques.

The effect of the frame

Because the picture is flat and relatively small, some strange effects can arise which influence our interpretation of the picture:

1. Subjects some distance apart can appear to be joined together – e.g. when a distant flag-pole 'grows' out of someone's head.
2. The apparent shapes and sizes of objects in a scene can vary with lens angle and distance.
3. Tone and colour in a shot can change the entire picture impact. Even a small area of brilliant color can dominate a shot if it is near the camera, yet be insignificant if the camera changes lens angle or viewpoint. Tones and colors can seem to vary against different backgrounds.
4. In the flat picture, quite unrelated subjects group together as a pattern, depending on the camera viewpoint.
5. The picture's frame seems to interact with nearby subjects, restricting or crushing down on them.
6. Where you position a subject in the frame can affect a picture's balance – whether it feels top-heavy or lop-sided.

Tight and loose framing

Tight screen-filling shots have a strong psychological impact, for the frame seems to confine and obstruct the subject.

Very loose framing leaves considerable space around the subject, which can produce an impression of isolation, emptiness, space.

Headroom

If there is too little room between the top of a person's head and the top of the frame, a shot looks cramped. Too much headroom and it looks bottom-heavy. Even a slight tilt of the camera alters the headroom considerably. The actual amount needed varies (the closer the shot, the less the headroom), but it should be consistent between similar types of shot.

Receiver masking

To provide as large a display as possible, TV receivers are usually adjusted so that the picture's edges are lost beyond the tube surround. If anything is too near the frame of your shot, therefore, it will be lost on most receivers. A camera viewfinder shows the *entire* picture, so try to keep all important subjects within a 10–15% safety margin around the frame edges.

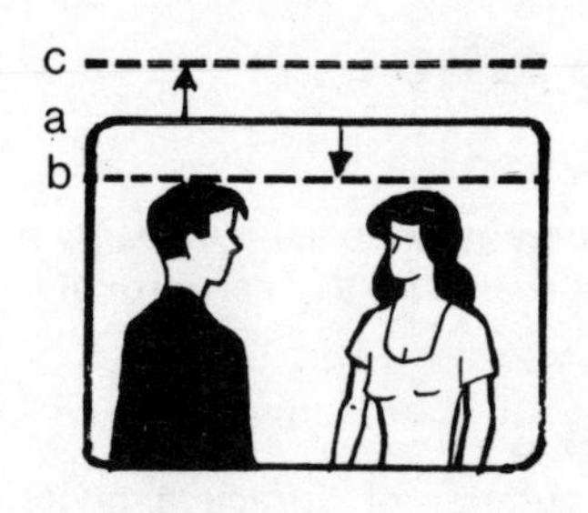

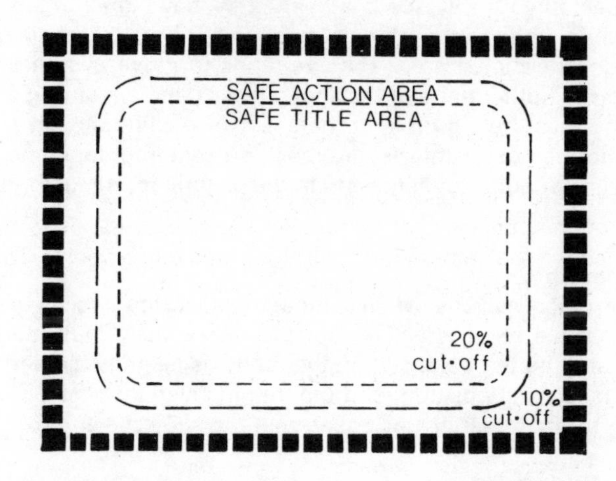

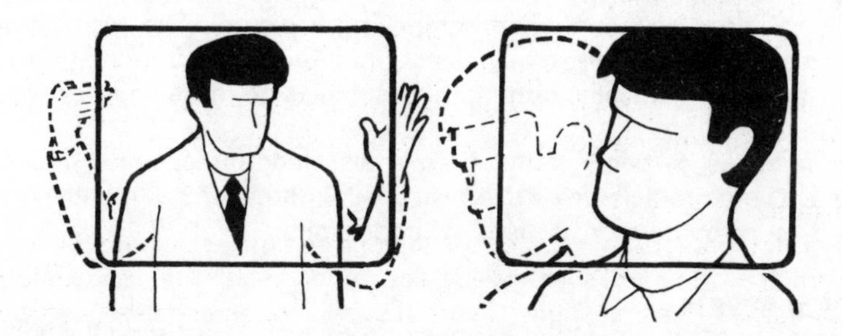

Headroom
The space between heads and the top of frame should be adjusted for attractive vertical balance, (a). It should not be excessive as in (c) which gives a bottom-heavy effect, nor too little as in (b) which gives a cramped effect.

Safety margins
Any subjects falling near picture borders are liable to be masked off by the TV receiver. So keep action and titles within the safe areas.

Tight frames
On very close shots, subjects can easily move outside the frame.

The way you frame a shot can change its entire effect.

Framing the Shot

When you are setting up any shot, keep an eagle eye on exactly how its frame interacts with the scene, for this can strongly influence the picture's effect.

Avoid routine centering
If you check over a number of attractive photographs, paintings or advertisements, you will see that very few have their main subject standing alone in the center of the screen. This is because it is usually the weakest concentrational area. The eye tends to move away from exact picture center to other parts of the shot, unless there are strong compositional lines concentrating on it, or there is eye-catching movement, color or form there. Most subjects are best off-centered to some extent, balanced against other tonal masses in the picture, for a more interesting compositional effect.

Offset framing
On more formal occasions, when someone is speaking directly to camera, centering can be most effective. But pictures usually look much more attractive and better balanced if the body is slightly angled (e.g. a three-quarter frontal position) and the framing slightly offset. This offset or 'looking room' generally increases with the subject's angle.

The rule of thirds
Except for special effects, it is best to avoid splitting the screen into equal, evenly balanced sections when composing a picture. The result is too mechanical. To avoid this, many cameramen divide the frame into thirds horizontally and vertically, putting main subjects on these lines, or where they cross.

The result is certainly better than a bisected frame, but it too can become a very predictable device. Using a 2:3 (fifths) or 3:5 (eighths) ratio instead can provide more satisfying proportions.

Framing people
If you are not careful, people can appear to sit, lean or stand on the border of the picture. So look out for this situation. It can be distracting – even ridiculous. Check too, that the frame does not cut the body and limbs at natural joins (neck, knees, elbows), for this too draws attention to itself. Instead, arrange framing at intermediate points, as in the basic shots on page 48.

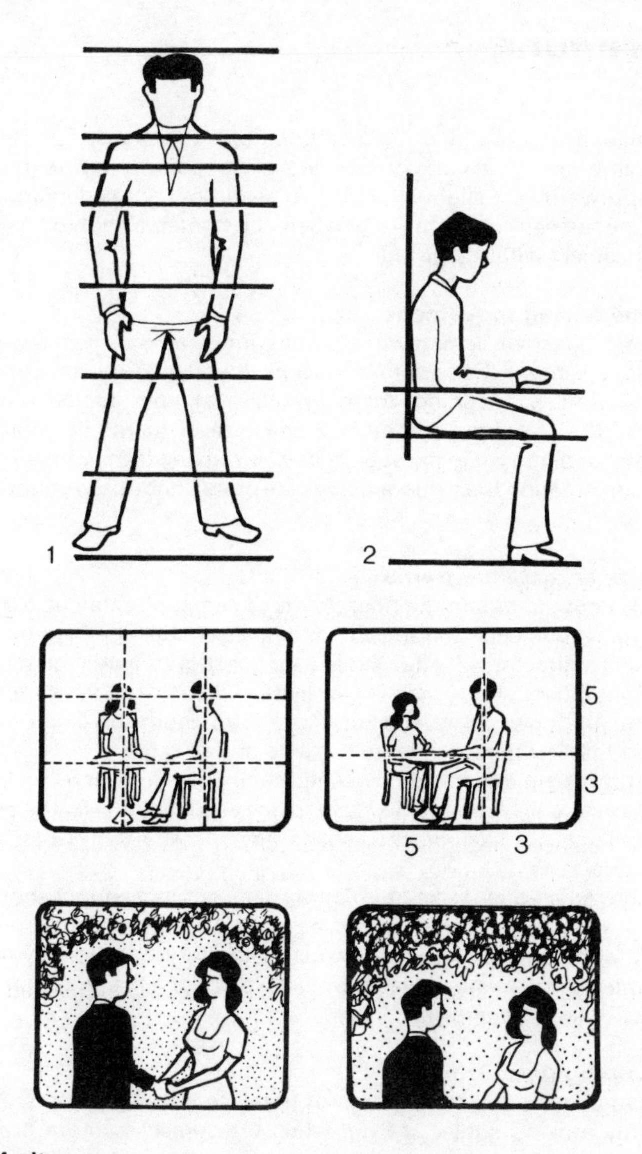

Framing faults
Avoid (1) letting the frame cut people at natural joints. Choose intermediate places, as shown here. Avoid them touching the screen edges (2).

Proportions
Dividing the screen into thirds produces standard, uninteresting proportions. Ratios of 3:5 or 2:3 give a more pleasing balance.

Ambiguous framing
If too prominent, a nearby subject may appear important. Is there someone hiding in the tree?

Unobtrusive reframing is a hallmark of skilled camerawork.

Reframing

As people move around, or the point of interest changes, you often need to reframe the shot to compensate. This adjustment might be as straightforward as a slight alteration to headroom, or as demanding as a simultaneous pan-tilt-zoom – e.g. when you tighten a shot to concentrate on one subject within a group.

Reframe during movement
The most effective camerawork always looks as if it has arisen quite naturally through action within the scene. If you alter the shot on a *static* subject (by tilting, panning, zooming, dollying), your audience becomes aware of the changes – perhaps over-aware of them! But make these same adjustments while the subject itself is *moving* (a movement as slight as a head turn) and the camera moves are quite unobtrusive, although still effective.

Subjects enter/leave frame
We come now to one of the finer points of camera treatment. Supposing someone who is speaking to camera in a close shot is joined by another person. The director will often cut to a longer shot showing them both. But this might interrupt the visual flow. Instead, he widens the shot to make room for the newcomer. The trick here is to zoom (or dolly) out while panning the first person over to the side of the frame.

A further technique for re-adjusting framing is often used to conclude a two-shot. Here the camera moves in to one speaker (usually the chairman or interviewer) to exclude the second person, who then exits when no longer in picture.

Whenever anyone walks out of a properly composed two-shot, part of the picture is left empty, so you need to reframe the shot smoothly, quickly, and decisively as the subject exits. Very occasionally you may deliberately hold an unbalanced shot after someone has exited, in order to emphasize their departure.

Maintaining good framing
We have already met the hazards of trying to hold tight shots of fast or randomly moving subjects. Even when the subject is more predictable (e.g. someone on a swing or a rocking-chair), the result is not particularly worthwhile. Frame-chasing, tight shots are usually tiring to watch, particularly if the subject keeps moving in and out of picture. It is better to widen the shot, to follow the movements more easily.

Adjustment
As people enter or leave the frame, you should normally readjust the framing.

Held frame
Sometimes a dramatic point is made by deliberately holding the frame still after an exit, to emphasize someone's departure.

Concentrating Attention

The more there is to see in a picture, the greater the opportunity for the audience's attention to wander. Although you might occasionally invite them to browse around at whatever catches their eye, more often you want them to look at certain features of the scene; to watch the action, not think about odd things in the background.

There are times when you should not let the audience see too much! If during a street interview, for example, they can read posters, see children waving to camera, passing traffic . . . they are bound to be distracted from the main subject.

Focusing interest

There are, of course, various ways of persuading your audience to look at a particular subject – a remark, a gesture, emphasized lighting. But the cameraman, too, can do a great deal to guide their interest:

1. By taking close shots, excluding other unwanted subjects.
2. By selecting plain backgrounds, or restricting depth (differential focusing) to isolate the subject.
3. By avoiding weak shots: side or rear views, high or distant shots.
4. By camera or zoom movement: dollying or zooming in on the subject; arcing round it.
5. By using composition to draw attention to the subject; e.g. isolating it, giving it prominence, adjusting picture balance, using converging compositional lines.
6. By careful framing, to keep distracting subjects out of shot.

Overdoing it

If you try to concentrate attention by filling the screen with detail, you may find that the over-enlarged picture loses its impact because it reveals the coarseness or crudity of design; or it loses clarity because there is no additional information to see (as in engravings); or the feature appears over-important relative to the complete subject (e.g. an enormously magnified detail that is normally overlooked).

Peeking through

Where you are having to shoot past foliage, mesh, bars etc., for a closer view of a subject, these obstructions may become a defocused blur, or may disappear altogether if kept close to the camera. Whenever possible, try to keep them out of shot altogether rather than rely on defocusing to make them unrecognizable, for even an indistinguishable blur degrades the overall image.

Foreground obstructions
Rather than including distracting foreground foliage, netting etc. in a shot, get closer and shoot through it. Even if you cannot avoid it altogether, this method makes it indistinct and less obtrusive.

Prominent subjects
Where a foreground subject is too prominent, better proportions are often achieved by shooting further away with a narrower lens angle.

Hiding distracting objects
By choosing the camera position carefully you can often hide distracting background objects behind people.

An illusion of depth enhances realism.

Composing in Depth

Although the TV screen presents a *flat* image of the three-dimensional world, an audience builds up impressions of depth and distance through various visual clues within the picture. It subconsciously interprets space by comparing relative sizes, converging lines (*linear perspective*), observing how one plane overlaps another (*masking*) and how the relative positions of subjects change as the camera moves (*parallactic movement*).

A picture that gives a strong feeling of depth is usually more convincing and more attractive. If few visual clues are given, it can be quite difficult to judge scale, space and distances.

Enhanced depth

There are several ways in which you can create a greater feeling of depth and realism in a picture:

1. Try to avoid having the subject isolated against a plain undetailed background, especially with flat diffused lighting shining from the camera's position.
2. If your shot contains familiar-sized objects (people, furniture), the viewer gets a better idea of distance and scale.
3. If the picture includes foreground planes (e.g. shooting past foliage or a window-frame) there is an enhanced illusion of depth.

Aim at natural effects

When you select or arrange foreground subjects to improve scenic depth, introduce them as naturally as possible. If you include foreground items too regularly or too obtrusively, they can draw attention to themselves. Resist the temptation to use them just 'to complete the picture'.

You should certainly steer clear of a 'peek-a-boo' camera style in which the subject is continually shot through grass, bead curtains, holes in a fence. . . . Effective enough when it *is* appropriate (e.g. the viewpoint of a person in hiding), this technique can be overdone.

So, too, can the gimmick of beginning a scene with a clearly focused but unimportant foreground object (e.g. a wayside flower), then pulling focus dramatically to blur it completely, bringing the real subject in the distance into sharp focus (e.g. a car on the highway).

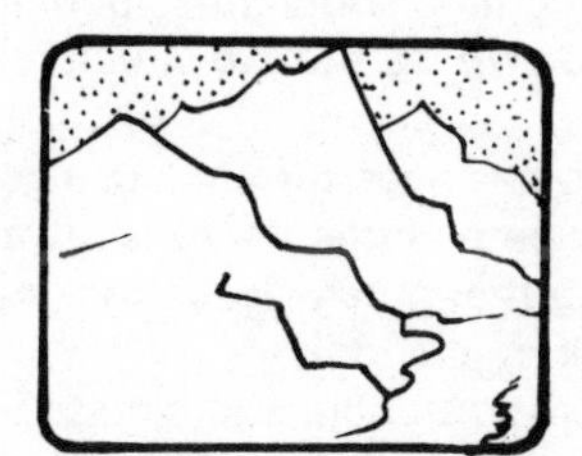

The illusion of depth
Foregrounds can help to create a greater impression of depth, distance and scale, particularly for isolated, remote subjects.

Foreground planes
Where the camera shoots the action beyond a foreground plane such as a window, screen, foliage, this can form a natural border to the picture.

Or 'Don't make a drama of it!'

Inappropriate Shots

You can see inappropriate shots of one kind or another every day on your home TV. Even professionals aren't infallible! Shots can be inappropriate in several ways:

1. The camera may not be showing the subject clearly (too distant, wrong angle, too close).
2. It may not show us what a person is referring to.
3. Composition may direct attention to the wrong subject.
4. Shots may be overdramatic, or miss visual opportunities.

Stylized camerawork

In the same way that one sees flat characterless *lighting* that merely illuminates but imparts no atmosphere to an occasion, so there is flat, characterless camerawork that provides routine shots from routine viewpoints, probably zooming in and out to bypass the labor (and skill) of actual camera movements.

On the other hand, when a cameraman has been turning out a series of perfectly good, straightforward but routine pictures, there is always the temptation to 'improve' shots, by getting that much closer, looking for a new angle or giving situations a more dramatic look.

However, a 'great shot' can draw disproportionate attention to itself. It may overemphasize a particular point. It can create an entirely wrong ambience. In a cookery demonstration, for example, dramatic shots could make the audience hold its breath, waiting for the fry-pan to catch fire!

A low-angle shot may show a speaker as imposing or threatening. But what if he is only reading a weather forecast? A canted shot has the power to convey instability or madness. But introduce it into a demonstration of farm machinery and the shot impact is quite misplaced. It means nothing and serves only to puzzle the audience.

Mannered treatment

Another bogus type of camerawork is the 'significantly composed' shot, where a prominent foreground object has been introduced to make a strong composition. It dominates the picture, although it should really only be incidental. Mannered treatment of this sort soon palls!

A good cameraman interprets the director's ideas, rather than pushes his own. If the director is inexperienced, most cameramen soon judge when to follow instructions and when to suggest alternative approaches! If a director insists on the unconventional, that is his responsibility.

Overdramatizing
Dramatic shots are fine, provided they are used at appropriate times.

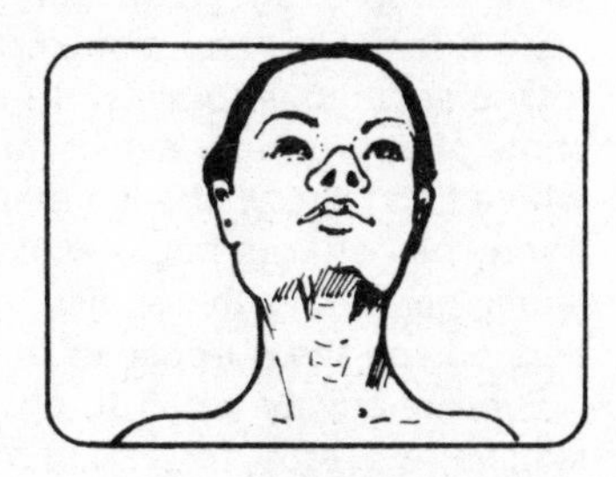

Contrived shots
'Significantly composed' shots too easily look bogus and simply draw attention to their own cleverness.

Difficult Shots

The cameraman is regularly faced with problems requiring split-second decisions. Some he just has to accept and cope with as best he can. Others he can often resolve by tackling them in a different way.

The wrong shape

What do you do when using your camera's horizontal four-by-three picture format to shoot subjects that are tall, round, long . . . ?

One solution is to show an overall view of the subject, then take close shots of appropriate detail. If the subject is very large a long shot of the whole thing shows the general form, but important features are not clear. It may be better to begin with an end-on viewpoint looking along it, then either pan along the subject or shoot it in a series of segments. Much depends on the purpose of the shot.

Even relatively small foreground objects can obscure a large distant subject. A nearby flower can blot out a far mountain. If you want to show the architectural form of a skyscraper, a distant shot would reveal its outline successfully – provided you can get far enough away, without intervening objects causing obstructions. A dramatic solution is to shoot up at the skyscraper from ground level. With foreshortening and strong perspective, the powerful effect provides limited information, but an arresting picture.

Spread subjects

Where subjects are spread around, you may have some difficulty in bringing them together in the same shot. Even an apparently simple situation, such as people at each end of a long table, illustrates this dilemma. Avoid panning between each subject ('hosepiping') over empty space, or just intercutting individual close-ups without a wide shot showing their relative positions. Oblique or end-on shots (perhaps over-shoulder viewpoints) can help the viewer here to maintain orientation.

Shooting into light

Apart from special effects (glare, flares, silhouettes) it is best to avoid shooting into strong lights. Not only does this set the video picture down, reducing auto-exposure and reproducing the subject as a silhouette, but the excess light may 'burn' on the camera tube, leaving an after-image and degrading picture quality. An extended lens sun shade (lens hood) may help, or even raising the camera and tilting down.

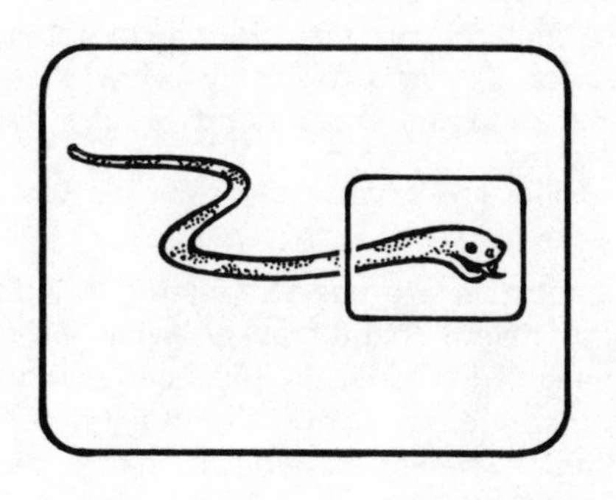

The wrong shape
Many subjects do not fit the screen's four by three format at all well. They are too tall, wide, long, or the wrong shape. Such subjects have to be shot in a wide view and then in localised segments.

Spread subjects
When subjects are spaced wide apart, you can often group them together in a well-balanced shot by using carefully selected viewpoints.

Changing Camera Height

The height-range and adjustment of camera mountings vary with the type of unit you use. Although you may very occasionally need extreme heights, such as a worm's-eye or bird's-eye view, steep upward-angled shots, high viewpoints etc., most situations require only modest height variations.

Pedestals (page 20)
Several kinds of pedestals are used regularly in studios. Some have pre-adjusted column height (hand-cranked while off-shot). Others use lightweight pneumatic or hydraulically balanced columns that are easily readjusted, while shooting, by a ring around the column (which also steers the mounting's wheels). A typical height range is 0.9–1.8 m (3–6 ft).

The columns of these pedestals need to be vertically balanced to suit the weight of the camera head and any accessories (e.g. prompter), to ensure that height adjustment is smooth and easy. This may mean changing counter-balance weights or gas pressure. Once this is done, you can raise/lower the camera slowly with precision or dramatically fast, taking care not to hit the top/bottom limits and jar the camera.

Why alter camera height?
The camera height you select usually depends on whether you want a natural or dramatic effect. When you are shooting people, camera height is normally around eye-level, varying according to whether the subjects are standing or sitting (1.2–1.8 m/4–5 ft standing, 1.1 m/3.5 ft sitting). In a dramatic situation you might shoot down or up at the subject, but certainly avoid doing so for everyday situations!

The base of a mounting often limits how near it can be placed to a subject on a platform (parallel, rostrum). A pedestal camera has little or no front overhang (unlike a camera crane), so it cannot reach over the subject but must stand back e.g. 1 m/3 ft.

Unexpected effects
When shooting with a camera crane, a strange effect can occur if you change height and tilt the camera at the same time, while holding the subject center-frame. The floor appears to tilt! Raising the camera and tilting down simultaneously makes the distant floor look as if it is tilting forward. Conversely, lowering the camera while tilting upwards produces a tilt-away effect. The illusion is more pronounced the greater the height change.

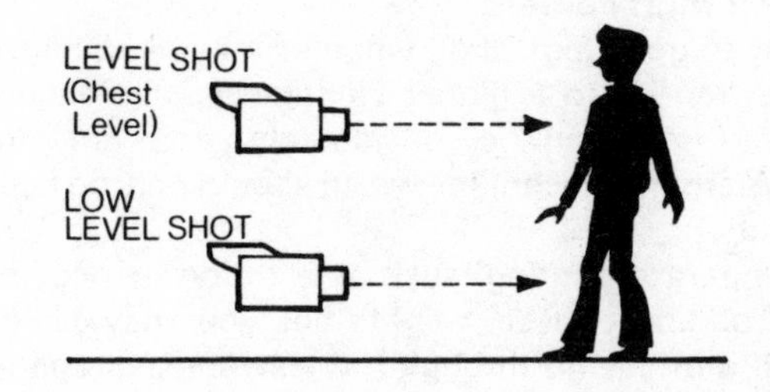

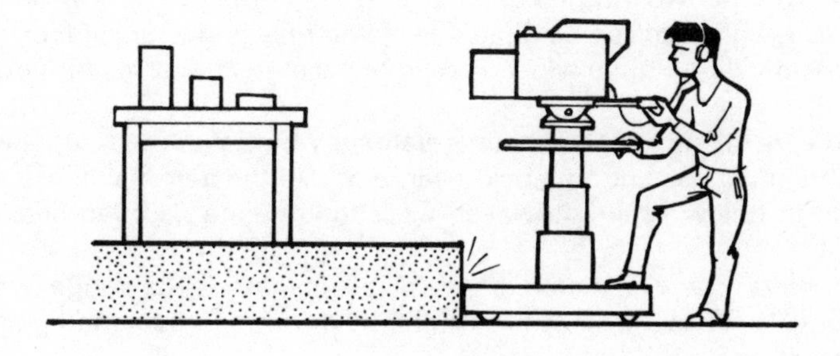

Level shots
The 'normal' camera viewpoint is usually around chest height. For somebody standing, this is about 1.2–1.8m (4–5ft) from the floor. For a seated person, the camera is typically 1.1m (3.5ft) high.

Obstructions
Platforms (rostra), steps and uneven ground impede most camera dollies. Although the jib of a crane can stretch over such obstructions, a pedestal or rolling tripod may, for example, be unable to approach and look down on a table.

High shots are not easy to control on simpler types of mountings.

High Shots

High shots provide both artistic and practical opportunities. A high-angle shot can make a subject appear weak, unimportant, inferior. It also enables the camera to see over things that would otherwise obstruct the shot.

Working with a high camera

The easiest way to get a high shot, when using a hand-held camera, is to move your position up to a higher vantage point. Holding the camera above your head is not only a considerable strain but produces shaky shots. It is an emergency technique which should only be used when there is no other solution.

A suitable camera mounting, such as a tripod or pedestal, holds the camera firmly at any chosen height; but you may have difficulty in operating it there or seeing through the viewfinder properly. In general, *pedestal cameras* are not easy to operate at maximum height, as you have to reach up to controls that have been arranged for comfortable operation at more normal working heights. Even if the viewfinder is well shielded from stray light and can be tilted down, you may find yourself looking up towards dazzling lights, which means the screen is that much harder to see.

If you are taking a high shot on a stationary pedestal, you might be able to stand on its base or on a box (riser, block). If the pedestal has to move around to follow action, however, a second operator is often needed to push it.

A *camera crane* can take shots at all heights in its range without difficulty. When working at full height, however, overhanging lights and scenery become a hazard.

Problems with high shots

As you raise the camera viewpoint, more and more of the floor is shown. As the human eye is accustomed to seeing a foreshortened view of space from a level camera, the floor area can look disproportionately large on picture, especially in interior shots. In many situations a high shot looks less imposing than a lower viewpoint.

If a camera is high and close, it can be very difficult to avoid shadowing the subject.

Finally, remember that high viewpoints often give unflattering shots of people (page 52).

Terms
High-angle shots tend to reduce the strength of a subject but give a clearer view of the action. They may be shot from an elevated camera or via a slung mirror or from a raised vantage point – e.g. a platform or tower.

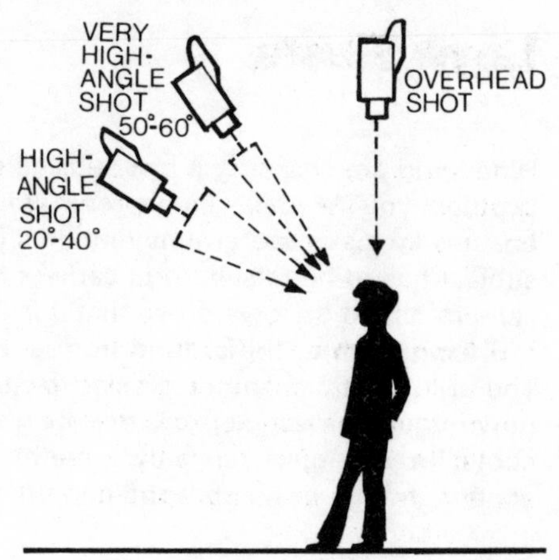

The effects of height changes
As the camera moves to a higher viewpoint, the floor becomes more prominent and the audience often feels less involved in the action.

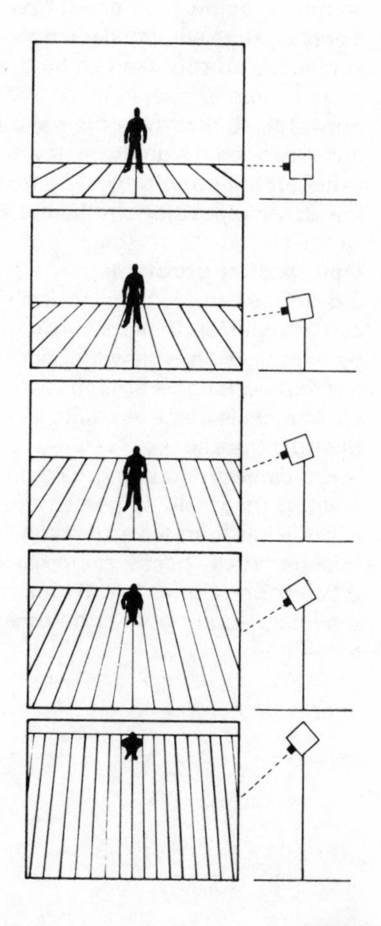

Often difficult but dramatically arresting.

Low Shots

When you are shooting a low subject, or one that is below your camera position, you inevitably shoot down at it. The result can look quite natural, but it may introduce an unintentionally dramatic effect. If so, either the subject has to be raised up to camera height (e.g. put on a table), or the camera has to be lowered so that it is level with the subject.

Shooting along the ground from a low-level viewpoint (ground shot), you will get much more striking pictures of low subjects than with a down-tilted camera at a normal height. But before selecting such low shots for a subject, consider whether this 'dog's-eye' viewpoint may appear strange or inappropriate in the context of the program.

The effect of low viewpoints

Although most subjects tend to look more impressive or important when shot from below the normal eye-line, take care, for some subjects take on a very strange and uncharacteristic appearance if shot this way! From low angles the surroundings may dominate and even overwhelm a subject.

Low shots of people tend to make them appear imposing, threatening, powerful, so that even their simplest actions – a glance, an outstretched arm – can seem significant. If you are using a wide-angle lens, this effect is exaggerated even further. So from an artistic point of view, always select low camera positions with care.

Operational problems

Even if the camera does not have to move, it can be very tiring to operate camera controls while crouching or kneeling for any length of time. It may be hard to see the viewfinder properly and to maintain good composition, particularly if lights are reflected in the viewfinder. A further hazard of low viewpoints is that even quite low furniture (chairs, table, stools) can easily obstruct the shot.

If a camera mounting cannot be lowered enough for the picture you want, it may help to use a mirror or a periscope attachment. But if the camera has to move around on shot, some sort of low-angle mounting or 'creeper' really becomes necessary. Its height may be variable from, say, 0.1 to 0.3 m/4 in to 2 ft. With a hand-held camera, even a castered board may help you to obtain the occasional low shot!

Terms
You can get low-angle viewpoints by using a low camera position, by placing the subject higher than the camera or by shooting via a floor mirror.

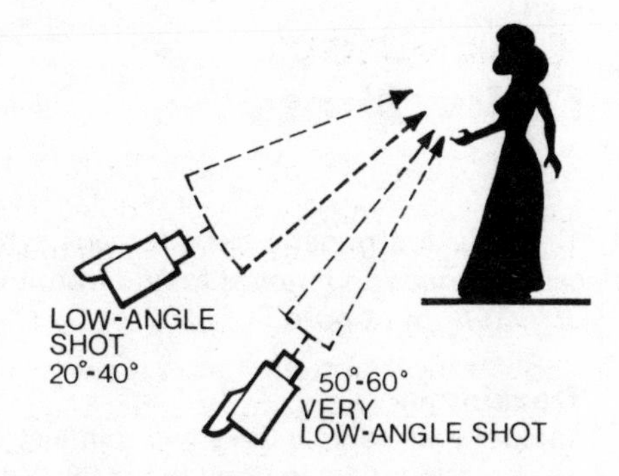

Impact
Low-angle shots make most subjects appear stronger and more imposing.

Low-angle dolly
Specially designed dollies are sometimes used when shooting continuous action from a low angle.

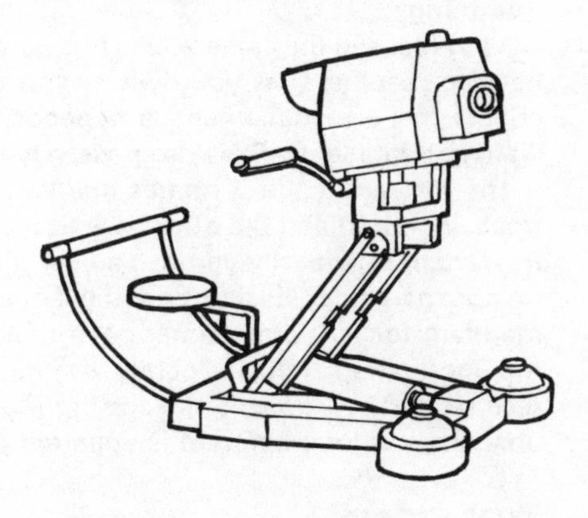

Pushing a pedestal with one hand, focusing with the other, adjusting height, maintaining good composition, yet not hitting anything on the way, is an acquired skill!

Dolly Shots

'Dolly' is the generic term for any wheeled camera mounting. As the camera mounting moves to and from the subject, we *dolly in (track in)* or *dolly out (track out).*

Tracking lines
The dolly wheels can be steered in a straight tracking line, a curved track, or an arc round the subject (page 88). Where camera moves are critical, its path may be marked on the floor or strategic floor-marks crayoned at intervals.

The effect of dollying
Unlike the action of *zooming,* which simply expands/contracts the same image, dolly shots take us within the scene, passing various objects as we move. We see the interaction of planes as the camera passes them, creating a strong illusion of depth and space.

Focusing
Having focused the camera lens at a particular distance, you will naturally have to readjust it as you dolly nearer or further from the subject. How critical focus readjustment is depends on the depth of field available, which of course changes progressively with distance.

The focusing control rotates one way to focus away from the camera (*focusing back*) and the other way to focus towards the camera (*focusing forwards*). How easily you can adjust the control while dollying depends on the particular design and the amount of compensation needed to maintain focus. Some cameramen make continuous adjustments while moving; others refocus gently as the main subject in the viewfinder picture starts to 'go soft' (for the high-definition viewfinder should reveal unsharpness long before it is apparent on a TV receiver).

Floor surface
The floor of a TV studio is normally specially laid to provide a flat, smooth surface, over which cameras can dolly freely without risk of picture bounce. On location, where floor irregularities can cause problems, the camera may have to remain stationary and rely on zooming to simulate dollying (moving around to new viewpoints while off-shot). Otherwise, special lightweight dollies fitted with pneumatic tires (instead of the usual solid-tired wheels) are used for limited movement. Occasionally, special floor rails may be necessary over uneven ground.

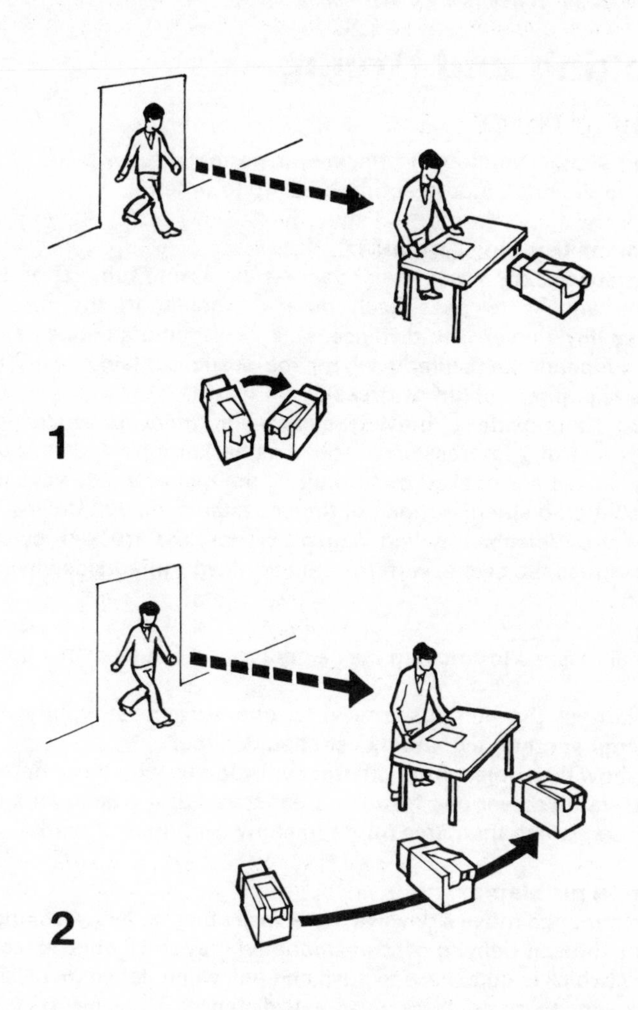

Fixed and moving viewpoints
When the camera *pans* around from a fixed position, the audience turn their heads to follow the action. The director may then cut to a new viewpoint to watch its continuation.
When the camera *moves* around, it changes the audience's viewpoint, 'walking them over' to the new position in a smooth continuous movement.

Trucking and Arcing

Although simple enough in principle, it normally takes a lot of patient practice to provide smooth accurate *arcing* to order.

Methods of trucking (crabbing)

As a camera *trucks (crabs)* straight across the scene, subjects at different distances appear to pass each other – rapidly in the foreground, proportionally slower with distance. This displacement creates a forceful illusion of depth, particularly where the scene contains many vertical features (e.g. posts, columns, trees).

Trucking alongside a moving subject (*a traveling or travel shot*) produces a strong impression of speed as background details slide past.

Some dollies are trucked by turning all their wheels sideways (as with the parallel/crab steering mode of the pedestal, page 20). Dollies steered by one set of wheels only (e.g. camera cranes) are 'trucked' by dollying straight across the scene, with the camera head turned sideways.

Arcing

The usual reasons for moving the camera round a subject in a tight circle are:

1. To correct the composition when one subject is slightly masking (obscuring) another, e.g. in an over-shoulder shot.
2. To show the subject from different viewpoints, e.g. the camera moves round a statue as various features are discussed; a craftsman speaks to the camera, which then arcs round to show his hands at work.

Operating problems

Some mountings move sideways more easily than others. A *rolling tripod* (at best a difficult dolly to control smoothly) may truck quite erratically. A *pedestal* (which is quite hard to push and pull when dollying/tracking) can also be very tiring to truck over any distance. The help of a second operator (tracker) may be necessary, to allow the cameraman to concentrate on focusing and composing the picture. For a trucking pedestal working at maximum or minimum height (i.e. fully elevated or depressed), assistance is essential.

Larger mountings such as *cranes* need a fair amount of space and time (depending on their size and design) to reposition from a normal in/out tracking line, to truck across the scene. A really tight arc may be impracticable on some types of mounting.

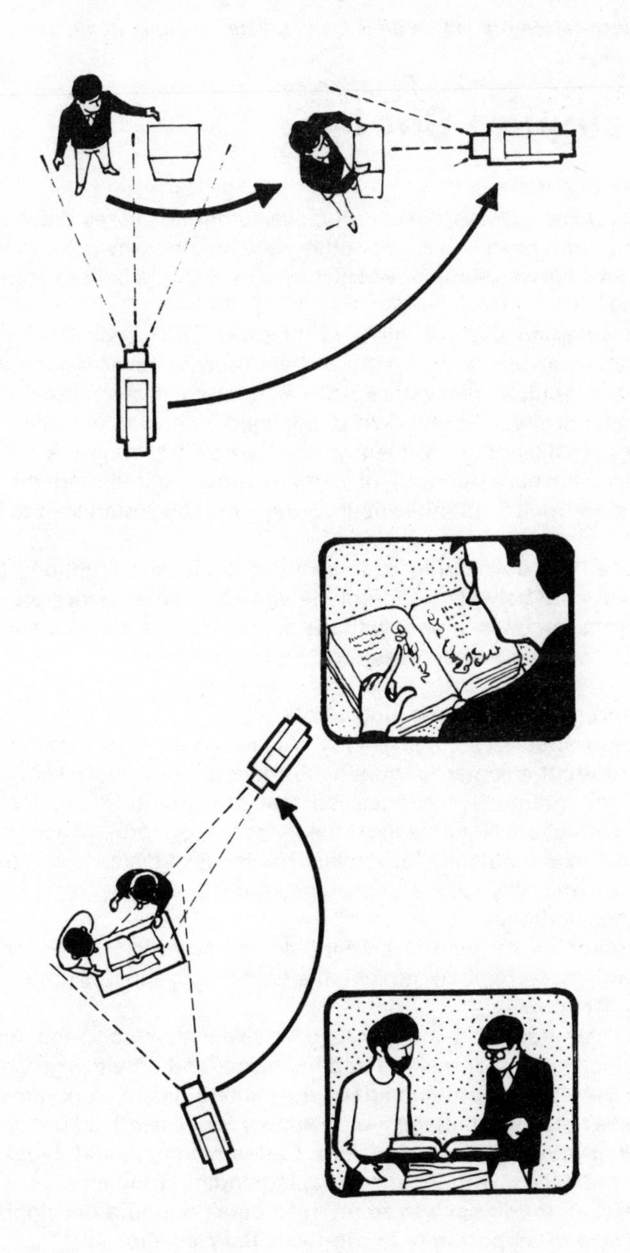

Moving subject
As the person moves to a new position, the camera arcs with him. The speed of the broad arc depends on how fast he moves.

Stationary subject
Here, having listened to the introductory conversation, the camera arcs round to see the subject itself more closely. The speed of this tight arc depends on the pace of the program.

But the more elaborate the treatment, the greater the opportunity for distracting mishaps to occur.

Developing Shots

In a *developing (development) shot,* the camera explores the scene as it moves on from one subject to another, or from one viewpoint to the next. In a smooth-flowing sequence of pictures, the camera builds up an illusion of space.

The developing shot can be used for several purposes. As the camera searches, it can show us where we are, slowly revealing details. The camera can gradually introduce us to a situation. It can compare a series of subjects, or demonstrate how one subject relates to another.

This type of shot is frequently used when the mood is solemn or romantic. The pace (tempo) of camera movement during the shot is usually slow, for it is intended to encourage growing interest or to build up tension.

The use of a developing shot avoids the visual disruptions that can occur with cuts between different viewpoints. The audience also feels a greater sense of involvement with the action, instead of their usual role as onlookers watching a succession of intercut fragments.

The mechanics of developing shots

Developing shots can involve very skilled, carefully controlled camera operation. As the camera mounting is moved (dollying, trucking, arcing) the picture continually changes, so that both framing and focus need smooth correction. Not only must the camera move around accurately but it must avoid any potential obstacles. The focused distance and the depth of field will probably vary, and the lens angle too may need adjustment to simplify dolly moves.

Developing shots need to be convincing and unobtrusive; indecision, hesitation, uncoordinated movements or focusing slip-ups draw attention to the mechanics.

Some cameramen prefer to use a relatively wide lens angle for developing shots, as it makes handling easier and provides greater depth of field. But under these conditions the camera has to work rather closer (otherwise the subject image would appear too distant), so distortion and shadowing may become a problem. Certainly you should avoid using a narrow-angle lens, with its unavoidable handling problems.

Remember, if you have to zoom in to detail during a developing shot, there is seldom opportunity to pre-focus the close-up.

Finally, try to use the same kind of dolly movement throughout, as any change of mode (e.g. from dolly to truck) is likely to cause the picture to bounce.

The varying viewpoint
The camera changes its viewpoint in a continuous exploratory movement, showing various aspects of the action.

The principles are obvious enough. The challenge comes when you have to do it.

Camera Movement

You can move a camera mounting at speeds from an almost imperceptible 'creep' to a rapid dash across the studio floor. The movement's safeness, smoothness and accuracy depend on the dolly and your skill.

Lightweight mountings

If a camera mounting is ultra-lightweight (e.g. a flimsy rolling tripod), camera judder or swaying shots are not easily avoided when dollying, particularly if the camera is much higher or lower than eye-level or if you try to guide it at arm's length.

Heavy mountings

A heavy pedestal is correspondingly harder to start moving and to stop at the end of its track. Take care not to strain yourself (or injure others) or damage nearby furniture, walls, etc. Gentle foot pressure at the base can help to start a dolly move.

Some cameramen operate with the left pan-bar (panning handle) folded upright at its centre joint rather than straight out in a horizontal position. They can then push the mounting with less strain and greater control, with the fingers of the left hand round the 'thumb-twist' zoom control and the right hand on the focus knob (at the side of the camera head or the other pan-bar). Camera handling is easier for steep downward or upward tilts.

Points to watch

Unless you actually want unsteady pictures for dramatic effect (e.g. simulating a man jostled in a crowd), always control all camera movements carefully. Make them smooth, deliberate and unobtrusive.

During any camera (or subject) moves, continually check focus for maximum sharpness. Try to avoid having to make a series of drastic corrections as the subject goes beyond the focusing limits. If both subject and camera are moving (getting closer or further away), focus-following is correspondingly more complicated.

The speed of camera moves needs to be artistically appropriate too. Slow dollying can create gradually increasing interest. It is therefore most suitable for gentle, unobtrusive changes in serious, thoughtful situations. But slow moves can also appear tedious, boring or frustrating, taking a long time to make a point. Fast dollying may be dramatic or exciting . . . or become a distracting dash between two points with a hazardous pull-up at the end.

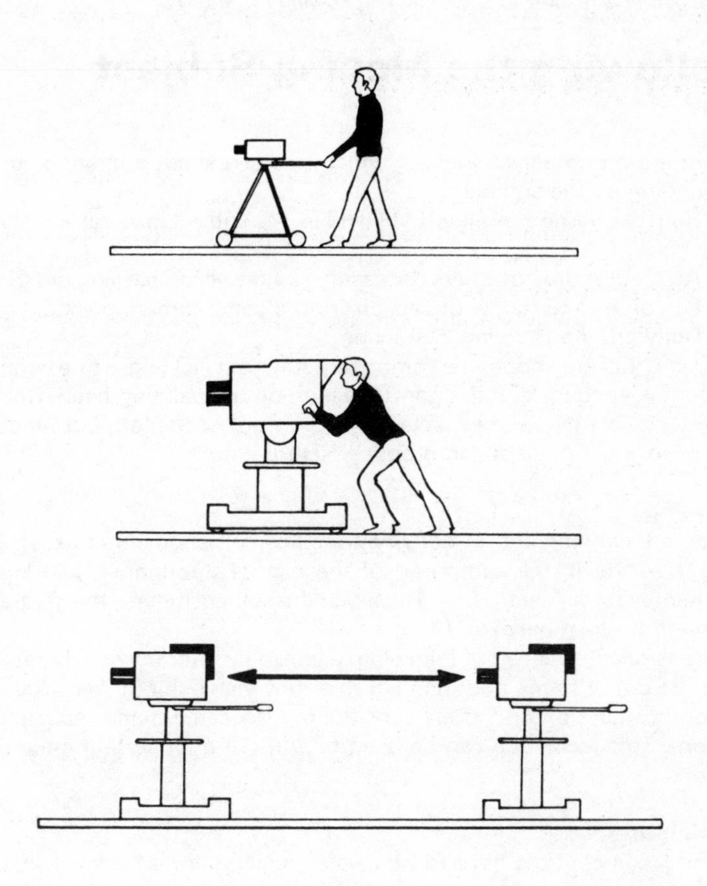

Lightweight mountings
Lightweight mountings move around so easily that it can be difficult to dolly them gently and accurately. They readily wander from a straight tracking line, vibrate, and shake the picture.
If a rolling tripod is flimsy and the camera elevated, it is best not to move on shot more than necessary.

Heavy mountings
Large studio cameras mounted on heavy-duty pedestals can be tedious to move around. They require a fair amount of effort to get them moving and to stop them, so it is important to take care.

Smooth movement
All camera movements should be smooth and controlled. It is all too easy to cause picture shake or jolt at the moment the dolly starts or stops moving.

Following the Moving Subject

There are a number of ways in which you can shoot a moving subject, depending on the situation:

1. Hold the frame completely still and let the subject move around within the shot.
2. As 1, but zoom out where necessary to save action passing out of shot.
3. Pan around to follow the action from a static camera position.
4. Dolly around to follow the action.

Unless you are shooting a completely static subject (e.g. a title card), it is seldom wise to lock off the pan/tilt action on the panning head. Not only may a subject move unexpectedly (and go out of frame!), but unlocking the head while on shot can produce picture jump.

Framing

If you are panning and/or dollying to follow somebody, try to keep them slightly offset in the same part of the picture throughout, just lagging behind the frame centre-line. The faster the subject moves, the greater the amount of offset needed.

Occasionally, instead of following action, a director may deliberately let it move out of frame and then cut to a new viewpoint (either taken on a second camera or shot after repositioning the camera and repeating the action). This technique can be used to edit out unimportant intervening action.

Anticipation

Some focus controls have to be rotated several times to cover their full focusing range, from the closest (MFD) to the far distance (infinity). Others are coarser to operate and need only a single turn to do the same thing. The type fitted to your camera affects how much adjustment you have to make to maintain focus.

Always be prepared for subject moves. Watch for the tell-tale hand and body movements that indicate someone is about to stand, lean back, sit, slouch, etc. Then you'll be ready to move with them smoothly and unobtrusively. Otherwise you are likely to be caught out and have to chase the shot to regain good composition.

Movement in closer shots

In longer shots a subject can move around quite a lot, yet still remain in frame. But as the shot tightens, it becomes harder to contain action and hold focus (page 58). Whether the shot size is adjusted to suit the subject's movements or the subject action is restricted to suit the shot depends on the situation.

Limited coverage
As the shot tightens, the amount of movement you can cover in this shot becomes more restricted.

Anticipating movement
Even experienced talent repositions without warning, or moves out of shot unexpectedly.

Shooting Graphics

Graphics have many regular applications in video production: for titles (title cards), lists, maps, charts, tables, insert stills, etc. They are supported on easels, music stands (caption stands), or attached to a scenic flat. Usually mounted on thick black card, sizes typically range from 30 × 23 cm (12 × 9 in) to 61 × 46 cm (24 × 18 in).

Don't take graphics for granted, and simply 'point and shoot'. A perfectly good graphic can easily appear on the screen as soft (unsharp), distorted, shaded, tilted, with distracting light patches. . . .

Lining up camera and graphics

Make sure that your camera is at right angles to the graphic, with the *lens axis* (center-line) in the middle of the graphics card. Check that the graphic is *level* (horizontal) and that it is not leaning back or sloping.

Check that your shot is appropriately framed. Is the complete graphic to fill the screen, or just a selected portion? When shooting a title card, is the lettering to appear as a *head-title, center-frame* or *sub-title?* It may not be obvious. Is your title to be combined with another camera's shot (e.g. one camera takes a map, another shoots the city names, and they are superimposed)?

Make sure that you are close enough to reveal all the required detail clearly – but not so close that wanted information at the edges is lost beyond the TV screen surround. Keep within the *'safe-title area'.*

If the graphic has a black background, the video engineer usually *sets the picture down (sit, bat down on blacks)* to ensure that an even black tone is reproduced. It helps initially if he *sets the video up* (graying the blacks) to enable you to frame the shot exactly.

Light problems

Light reflections or glare can easily obscure parts of a shiny graphic or glossy photograph. Sometimes slightly tilting the card or adjusting camera height clears the problem. Otherwise, dulling with a wax spray or re-lighting may be necessary.

Where you have to pan around a graphic in a tight shot, zooming is easier than dollying. But camera distance and lens angle may be a compromise between possible shadowing and coarse camera-handling.

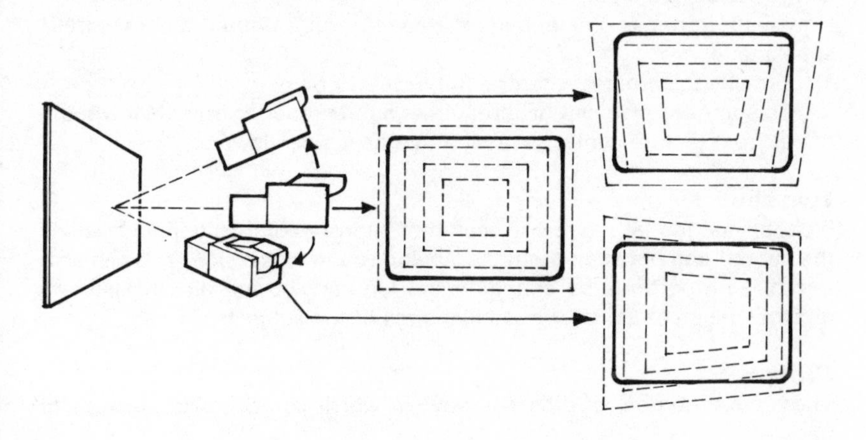

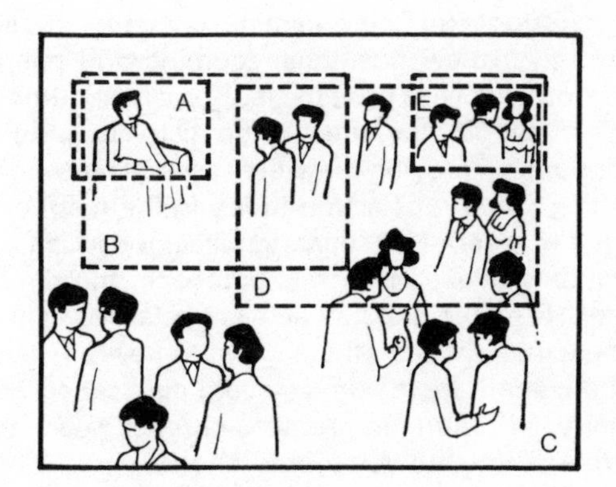

Title positions
Titling can be located in several positions in the frame.

Shooting graphics
Graphics become distorted if they are not shot from a straight-on central position. Keystone distortion occurs if the camera is off-centre.

Exploring graphics
The camera can sometimes 'pseudo-animate' graphics by shooting them in sections, zooming in on detail, zooming out to take in larger areas.

Shooting People

Most productions are mainly about people. So it is not surprising that shots of people are the staple concern of camera technique.

Single shot (page 50)
How you frame a single person depends on whether the person is speaking:
1. Directly to camera: a central full-face viewpoint.
2. To someone else, out of shot: either a three-quarter frontal viewpoint, offset from picture center, or a slightly offset side-view.

Two shot
If people are too far apart, you can often improve composition and reduce the central gap between them by moving round to the side. You can also improve proportions by changing the lens angle, and altering camera distance helps to adjust the relative sizes of the subjects.

Groups
There are a number of different ways in which you can shoot groups of people:
1. *A one-camera shoot.* (a) One camera continuously shoots all the action, arcing round to new positions, zooming and panning between individuals as unobtrusively as possible (e.g. as a head turns). Avoid using whip pans between people; the effect is too distracting. (b) The camera shoots discontinuously, moving between several viewpoints. It either selects parts of the continuous action (missing the rest), or the action is specially arranged in a series of separate takes or sequences.
2. *A multi-camera production.* In this approach, the director guides a group of cameras. He either intercuts between relatively static viewpoints, or shoots on one camera while others move to new positions.

When part of a camera team, aim to avoid duplicating other cameras' shots – particularly if shooting impromptu unrehearsed action. A nearby monitor can help to show other cameras' shots.

Crossing the line
If the camera viewpoint alters unexpectedly, the audience can lose their sense of direction. For example, when you are shooting two people in conversation, always imagine a line cutting across the floor between them. If, *while shooting,* you dolly and cross this line, your audience can follow the viewpoint change. But if you *switch* between camera positions either side of this line, subjects on left and right of the frame suddenly exchange positions! Successive viewpoints must therefore be chosen carefully to avoid *'crossing the line'* and causing audience disorientation.

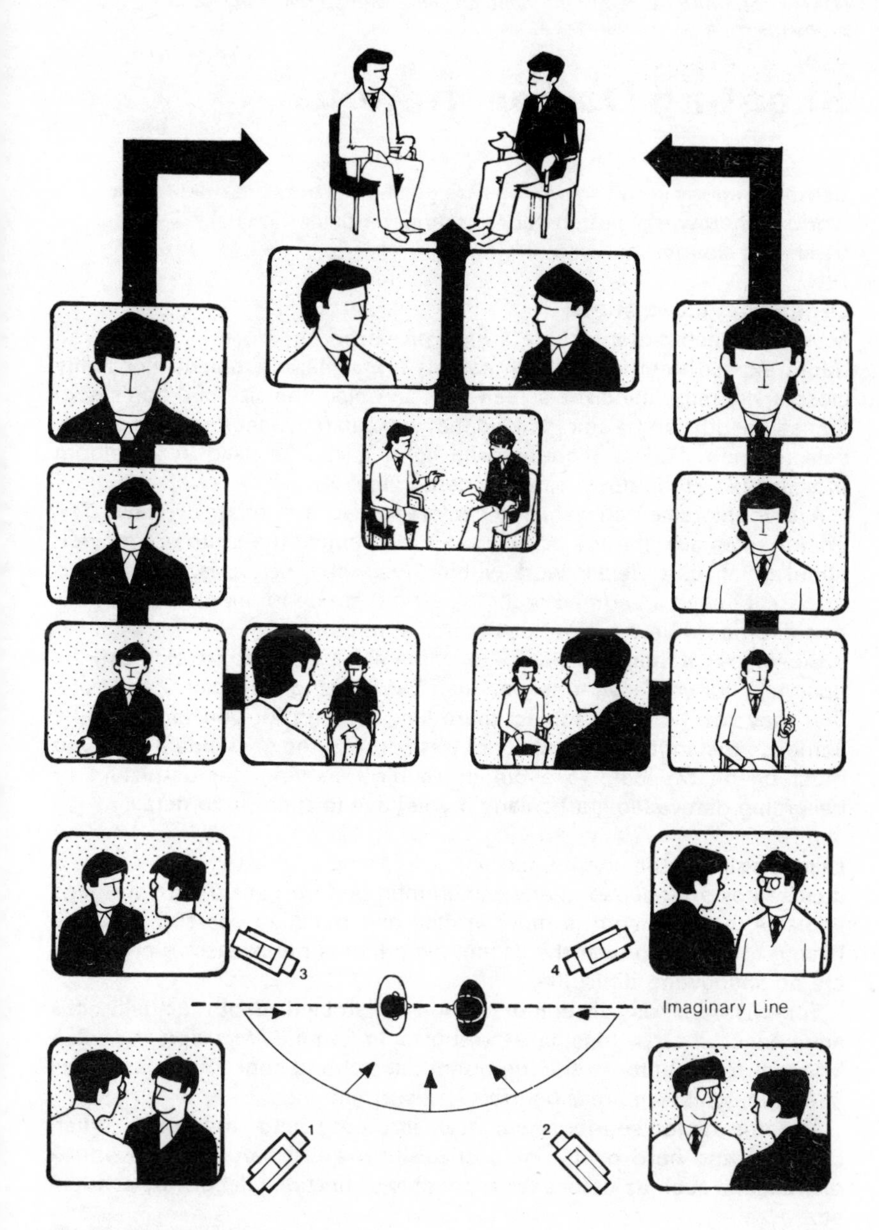

The formal interview
A number of 'standard' shots are used for most interview situations.

The 'imaginary line'
Shots can be intercut (cross-cut) between viewpoints *on the same side* of an imaginary line joining two people (1 & 2 or 3 & 4). But interswitching *across* the line causes jump cuts (1 & 3, 1 & 4, 2 & 3, 2 & 4). There are no problems of this kind when *dollying* over the line.

Demonstrations – whether of hobbies, handicrafts, scientific, sales – provide problems for the cameraman.

Shooting Demonstrations

Demonstrations show how things are used, how they are made, how they work. Camerawork is often very exacting, particularly where detail has to be shown clearly.

Organizing close shots
Most demonstrations include a high proportion of close shots. As you might expect, depth-of-field limitations are a major problem. For really close shots (e.g. filling the screen with an object the size of a hen's egg) focused depth can be so restricted that you can only focus sharply on one selected part. Unless high-intensity lighting can be used, it is seldom practicable to stop down to any extent (page 24).

A demonstrator can help the cameraman in a number of ways. The golden rules are: put the object on a pre-arranged mark; do not move it about or obscure detail; work within a very confined area. It is best to discuss items in an agreed order, offering them up to the camera in turn. Avoid jumping between various things at random.

Close-ups of items held up to camera by a demonstrator can be uncertain, for you have to rely on their hands being very steady and hope that they will hold the subject there long enough for you to focus and compose the shot. Snatch shots of this sort may be unavoidable. So you must be on the ball, to avoid the subject moving out of picture or becoming defocused, particularly if you have to zoom in to detail.

Lens angles
If close-ups are needed, many cameramen prefer to use a narrower lens angle. Although depth is more limited and handling coarser, it avoids having to work too near the action, distortion is not excessive and there are no shadowing difficulties.

For many kinds of demonstrations restricted depth actually has advantages. It helps to isolate the subject (*differential focusing,* page 30). You can focus hard on the important subject and concentrate audience attention, while suppressing irrelevant surroundings.

Try to avoid spurious subjects intruding into the shot. When concentrating hard on the subject itself, you can very easily overlook distractions such as defocused highlights, reflections, labels, bystanders etc.

Viewpoint
If the viewpoint is poor, important detail may not be visible.

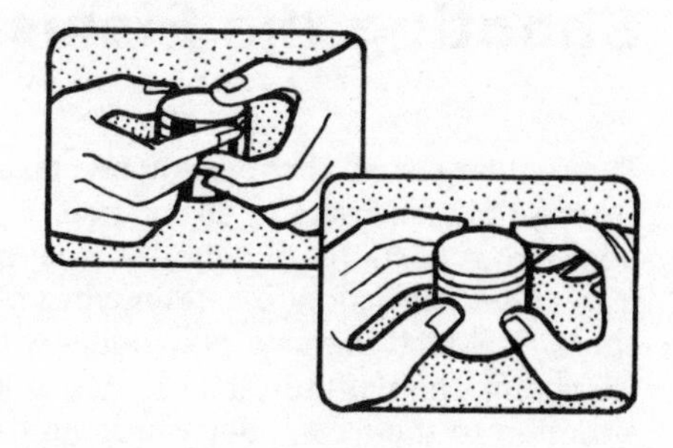

Table marks
If a demonstrator puts items down at random the camera may not have any good shot opportunities. Neat unobtrusive marks on the table can ensure that things are positioned accurately every time.

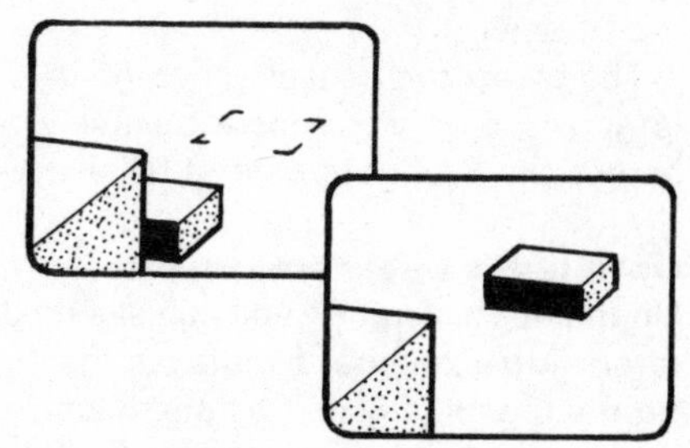

Steadying objects
If items are held in mid-air, pictures are likely to be unsteady and defocused.

Selective depth
Whether restricted depth of field helps (by isolating an individual item) or hinders (by not showing sufficient detail) depends on the purpose of the demonstration.

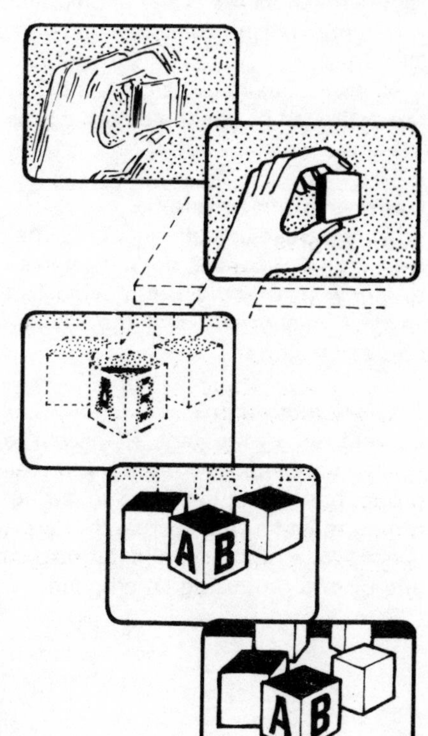

Shoot the pianist . . . in a fashion appropriate to his performance.

Shooting the Pianist

The regular range of shots you can take of a pianist and his instrument is surprisingly restricted. Of course, various 'novelty' approaches are possible: reflections, silhouettes, shadows or unusually angled shots can give added attraction to certain types of presentation. There are also trick shots, in which pianists play duets with themselves or appear in multi-viewpoint segmented pictures. But unusual treatments can easily draw attention to their own cleverness and away from the performer and the music itself.

The piano can be a challenging instrument to shoot successfully, for its large mass only frames attractively from certain directions. From other angles the keyboard cannot be seen – or even the pianist!

Optimum camera treatment

On the opposite page you can see the basic range of compositions a piano offers. Most keyboard shots put the pianist to the left of the picture, for the right hand usually carries the main melody. From the left-hand end of the keyboard, shots are pretty limited, for they do not generally intercut well with other viewpoints, and unless you are careful *reverse cuts* arise (page 98).

At normal camera heights, moving round the piano is confined to a rather limited arc, as the pianist's back, the piano lid or its body obscures the shot.

Depth-of-field problems

To obtain close-ups of fingerwork, the obvious solution is to use narrower lens angles. However, depth is then so limited that it may not be possible to follow focus on a fast-moving hand. Restricted depth, together with severe foreshortening of the keyboard, can make such detailed shots unpredictable.

Camera movement

Camera moves are generally keyed to the pace and mood of the music. Rapid shot-changes from a variety of positions may be great for a pop group, but for a quiet passage during a recital of classical music camera moves must be so slow that they are almost imperceptible. The audience is not aware of the change but responds to the effect, the blend of sound and picture producing an emotional whole.

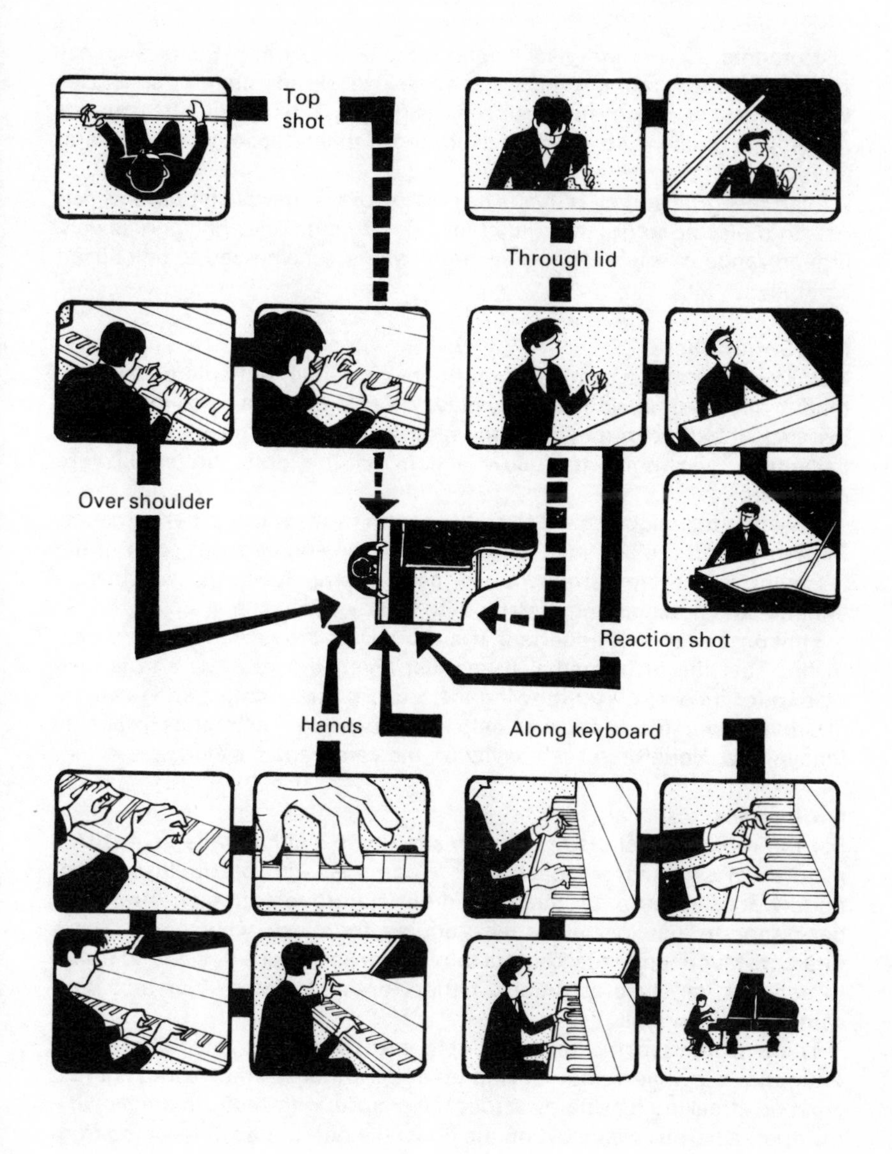

Shooting the pianist
When presenting a piano performance, the basic range of attractive shots is relatively limited.

Badly shot, an instrumental performance turns into a succession of frustrating puzzle pictures.

Shooting Instrumentalists

Instruments vary in the visual opportunities they offer. Some are best seen in a wider shot; others have very localized interest, so closer viewpoints are more successful. Attention is often split, the camera needing to follow two hands performing intricate actions (e.g. guitars, lutes).

Even this brief survey reminds us how diverse musical instruments can be, from piccolo to grand organ. Most instruments offer only a relatively limited range of interesting, meaningful shots, which need to be chosen carefully.

Instrumentalists

Instrumentalists in a seated group move very little, but solo performers tend to alter their position as they play. When shooting a solo violinist, for instance, the director may need to readjust the camera position or switch to another viewpoint to follow action, as the optimum angles are restricted.

Shooting techniques need to suit the form of the musical performance. Sometimes off-the-cuff inspiration conveys the spontaneous spirit of the occasion. For more formal music, careful preparation is essential to enhance the audience's understanding of the music.

Shot organization is important. It is too easy to arrive at a new shot, only to find that the instrumentalist has just finished a passage and is now resting for a dozen bars. Knowing the musical arrangement and having a good memory for shots will help. Although the director is probably following a score, he relies heavily on the cameraman's alertness.

Orchestras

For the most part, shots of an orchestra involve either wide generalized pictures of sections or closer shots of groups, with detail close-ups of fingerwork. Because of the spread-out nature of an orchestra, it is necessary to use narrower lens angles for close viewpoints. Some directors favour continual camera movement as the lens dollies and trucks to select a variety of positions. Others prefer to intercut or mix from semi-static positions.

There is always the temptation to take shots that are different or dramatic: a low viewpoint looking up at a clarinetist; the colored lighting creating streaking lens flares across the picture; reflections in the bell of a trumpet; a harpist's shadow on the floor; the old routine of focusing from harp strings through to the players beyond; rhythmical feet; swift zooming to audience reactions. Great ideas for suitable occasions, but make sure that they *are* appropriate!

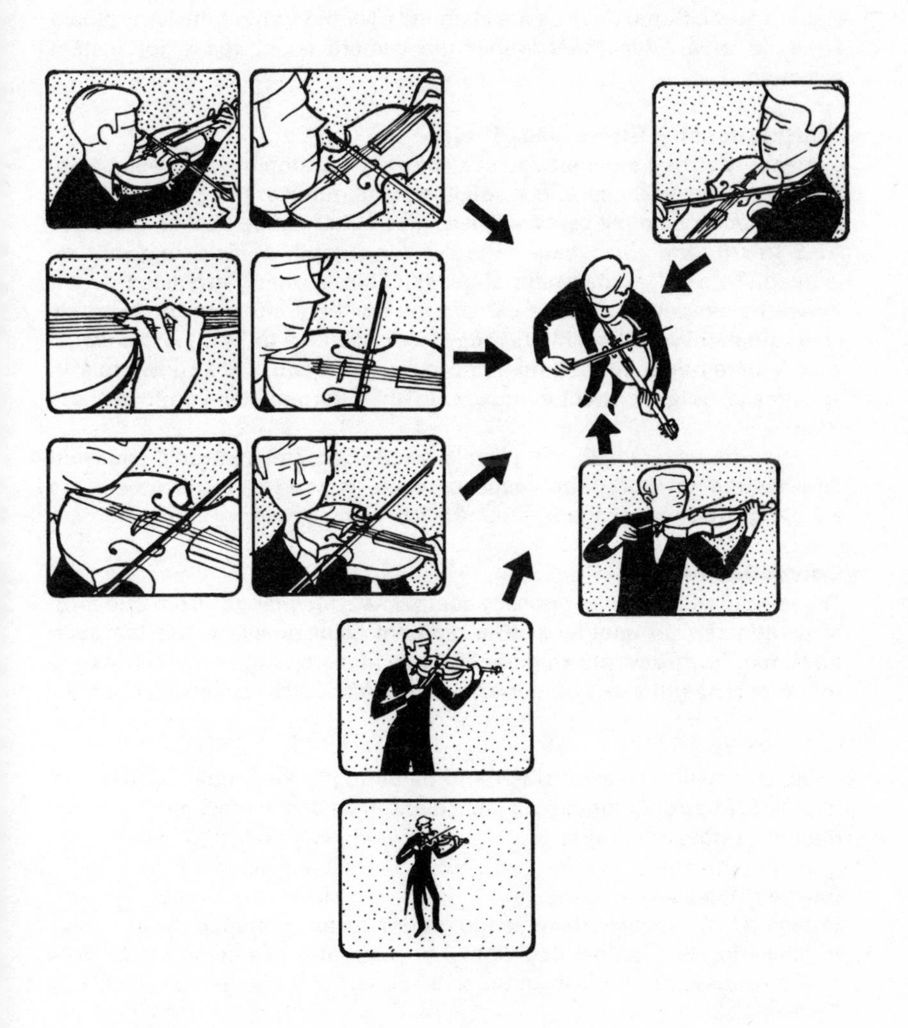

Shooting the violinist
The shape of any instrument, and how it is played, determines which angles provide the best shots.

Filters and Effects

Shooting through a filter or optical accessory can produce a variety of visual effects. Some devices are clamped over the front of the lens, others are held in a *filter-wheel* within the camera head, ready for instant selection.

Neutral density filters – ND filters

Under very strong sunlight you normally have to stop the lens well down, to prevent over-exposure. But small lens apertures are not always desirable, (a) because most camera lenses give better image quality at around *f*/5.6 to *f*/8, and (b) because the resulting depth of field may not be wanted. So a gray-tinted neutral-density filter is often introduced, to cut down the overall light to the camera tube without affecting the picture's color quality. Typical ND filters range from 10% (1⁄10) to 1% (1⁄100) *transmission.* Where bright lighting might normally force you to stop down to *f*/16 in order to obtain correct exposure, a 10% filter enables you to work at *f*/5.6.

Even under average lighting levels, an ND filter can be used if you want one camera to work with reduced depth (e.g. *f*/1.9) while others are shooting the same scene at around *f*/5.6, for example.

Corrective filters

These colored filters can optically compensate for changes from one form of illumination to another with a different color quality (color temperature), e.g. from daylight (high kelvins) to tungsten light (low kelvins). In many cameras this can be done by *auto-white balance* readjustment.

Star filters

Clear discs with closely scribed grid patterns produce multi-ray patterns (e.g. 4, 6, 8) around highlights (specular reflections, lamps, etc.). Turning the filter rotates the rays.

Image diffusion

Diffusion discs create effects ranging from slightly softened picture detail to dense fog with haloed highlights. You can also use nylon net or hose over the lens, or shoot through clear glass lightly smeared with oil or grease.

Image treatment

Multiple images can be produced optically by prismatic or faceted lenses, ribbed or stepped filters, or mirror kaleidoscopes. You can also *distort* the picture by shooting through ripple glass or via a flexible plastic mirror. Shooting through a mirror-plastic tube surrounds the central image with colorful reflections. *Inverter prisms* can tilt (cant), invert, turn the image sideways or rotate it.

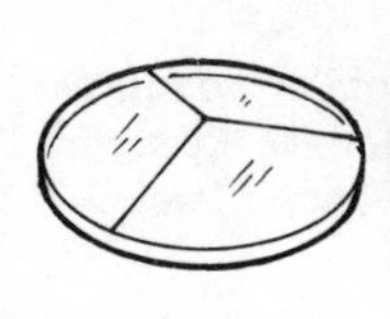

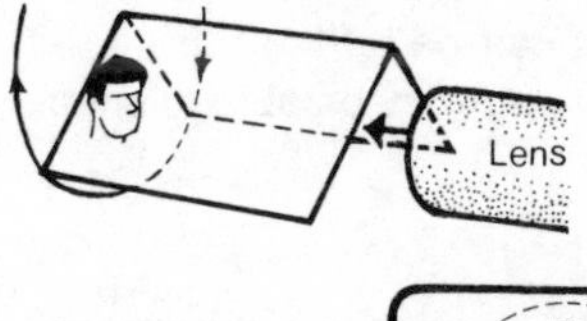

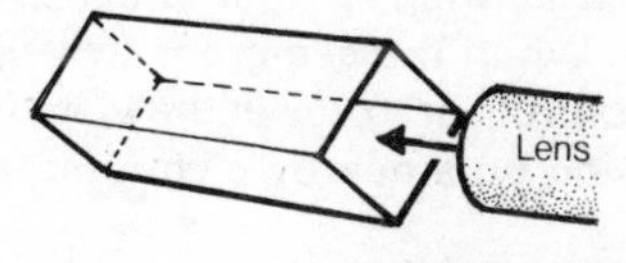

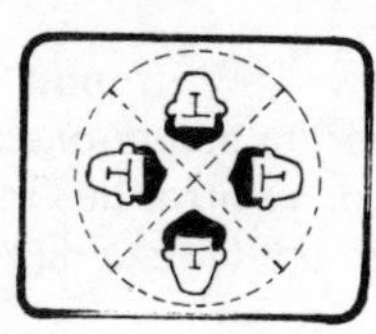

Prismatic lens
A multi-faceted lens provides several identical images which can be rotated by turning the device. In the second type, the central image remains still as the peripheral images move around it.

The kaleidoscope
Shooting through a three- or four-sided mirror tube produces an upright central image, surrounded by angled reflections.

All Done By Mirrors

At times you may want to shoot a subject from a very unusual angle:
1. To show details that are not normally visible.
2. To get round obstructions that are blocking the shot.
3. To achieve a dramatic effect.

This may involve an overhead (bird's-eye) view, an up-tilted (worm's-eye) viewpoint, a floor-level shot, or shooting from a good but inaccessible vantage point.

Although today's small video cameras are highly mobile, adaptable units, extreme shooting angles can still pose problems. More bulky studio cameras, with their larger zoom lens systems and viewfinders, are considerably less flexible.

The direct approach

You may be able to place the camera exactly where you need it, but for particularly hazardous positions (e.g. directly overhead or halfway up a wall) it is often quite impracticable to support camera and cameraman there. When the camera is awkwardly angled it can be difficult to adjust the controls or to see in the viewfinder, although you may be able to check focusing on a nearby picture monitor.

Using mirrors

By shooting via a mirror, you can effectively reach positions that would otherwise be impossible or require elaborate safety precautions.

Particularly where resources are limited, a mirror has many advantages and the camera can swing away in a moment to take direct shots. Typical mirror shots include:

1. Top shots of a demonstration table using an overhead slung mirror. The camera, at normal height, tilts up, shooting via the mirror.
2. A high-angle view of action that would otherwise require a camera crane.
3. Level shots of a subject high up on a wall (statues, balconies, windows).
4. Low-level and low-angle shots through a floor mirror or a periscope.

Problems with mirrors

Mirrors have their drawbacks. The picture is often reversed horizontally (laterally) or vertically unless corrected (electronically by *field/frame reversal,* or by interposing a second mirror).

Heavy glass mirrors often take time and skill to adjust, and can obstruct lighting. If the mirror is small or distant, its coverage may be too restricted. Ideally, mirrors need to be surface-silvered to avoid degraded double images, although lightweight mirror-surfaced plastic sheeting is quite satisfactory for some purposes.

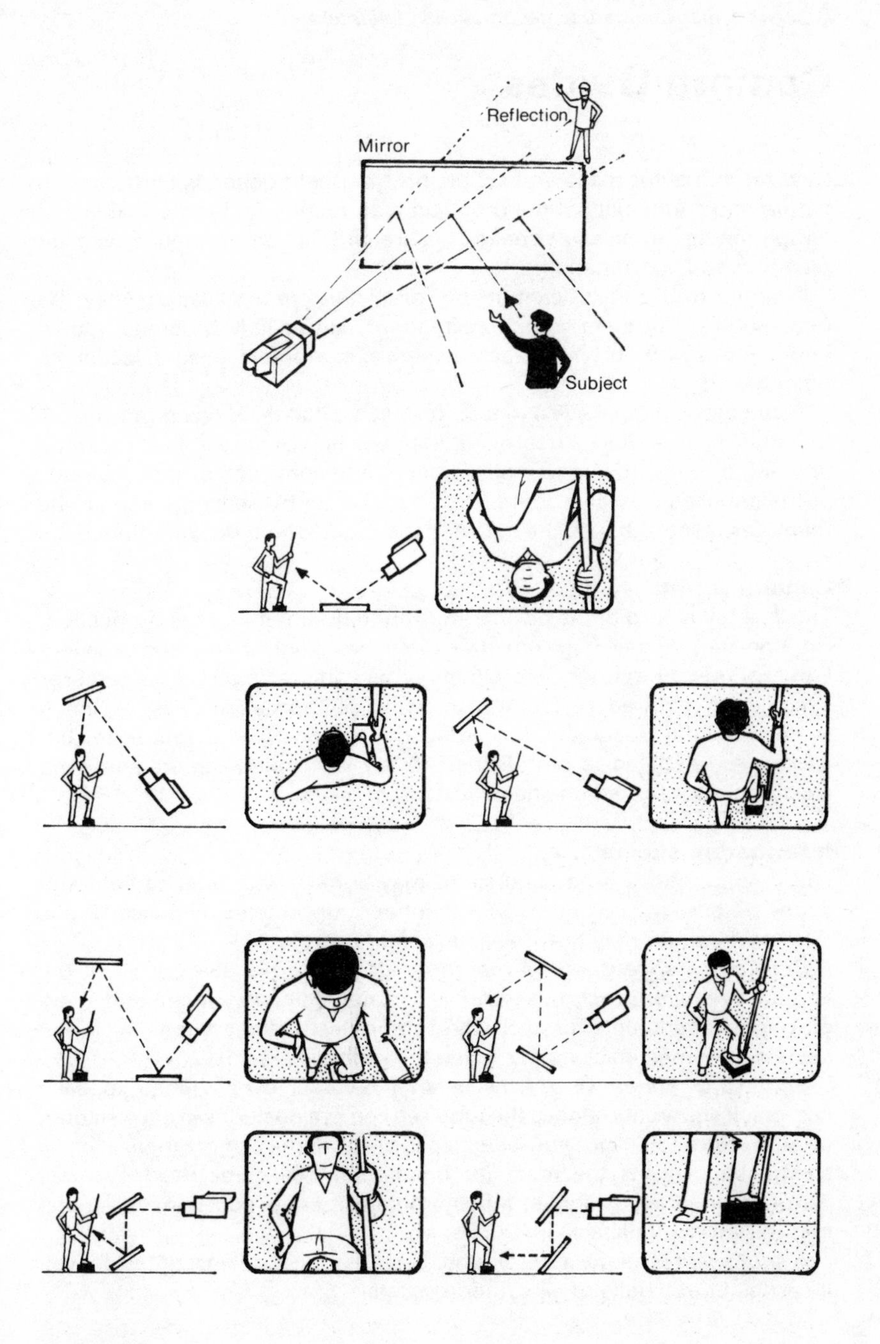

Basic principles
The reflected image is reversed laterally (side-to-side) and appears as far behind the mirror as the subject is in front. Focus on the subject image, not on the mirror's surface.

Camera Guides

How far a director plans and scripts his treatment depends partly on the nature and complexity of the production and partly on his own skills and temperament. Some shows need little preparation, others require detailed planning and coordination.

Where shot details cannot be predicted, improvisation may be unavoidable. But even complex situations can usually be broken down into a framework in which each camera has a prearranged selection of potential shots.

Some productions have a regular format, so that the director only needs to outline each camera's treatment: 'Camera 1 gives me a long shot for the opening of the show; mid-shot entrances and contestants, and close-ups of the panel as they contribute . . .'. Then it is up to the cameraman to be ready for these shots as they arise. Paperwork would be superfluous.

Camera script

This is a full record of the picture and sound treatment for the production. On one half of the page are the dialogue, stage instructions, action, lighting, sound treatment, etc. On the other half are details of the cameras used, shots required, picture transitions (cuts, mixes/dissolves, wipes). It is an essential document for production planning and reference, but usually too detailed to be followed by a preoccupied cameraman, who uses various information sheets instead.

Information sheets

For simpler shows an outline script may suffice, with brief camera and sound details, 'in and out' cuewords for announcements, film inserts, etc. (The cameraman may use this as a '*camera card*'.)

Breakdown sheet/show format/running order. This gives a list of the various events or program segments in order. It shows camera and audio pick-up arrangements for each, the setting used, talent names, etc. (The list is sometimes inaccurately called a *'rundown sheet'*.)

Shot card/camera card. This is an individual card clipped to each camera. It shows at a glance the required camera positions in each setting, its allocated shots (shot numbers, types of shot), the camera moves, basic action, etc. This is the main (or only) reference guide used by most cameramen in the course of a busy multi-camera production, to supplement intercom (talkback) instructions.

A *shot sheet* is useful as a treatment summary list for unscripted shows. It carries brief details of all cameras' shots.

RUNNING ORDER

Page	Scene	Shots	Cams/Booms	D/N	Cast
1	1. INT. WOODSHED	1 - 9	1A, 3A, A1	DAY	Mike Jane
3	EXT. WOODSHED	10 - 12	2A, F/P	DAY	Jane Jim
	RECORDING BREAK				
4	2. INT. SHOP	13 - 14	4A, B1	DAY	George
6	3. EXT. WOODS	15	1B, 3B, A2	DAY	Mike George
7	4. INT. SHOP	16 - 20	4A, B1	DAY	George

CAMERA CARD

CAMERA ONE — THE OLD MILL HOUSE — STUDIO B

SHOT	POSN.	LENS ANGLE*	SET
2	A	24°	1. WOODSHED LS TABLE PAN MIKE L. to window.
5	(A)	35°	MS MIKE moves R. to stove JANE into shot L. Hold 2-shot as they X to wood-pile.
	MOVE TO POSITION 'B' during Shot 6.		
21	B	10°	BCU Door latch. ZOOM OUT to MS as door opens.
24	(B)	24°	CU back MIKE'S head. As he looks up, TILT UP to MCU of PAT (POV shot)

* OPTIONAL

Multi-camera Production

Whenever a fairly complicated production is to be presented 'live' (transmitted as it happens), or 'live-on-tape' (recorded as a straight-through performance), there are considerable advantages in shooting the action with two or more cameras simultaneously.

To select between different video outputs, a *production switcher (vision mixer)* is used. This has switching and fader circuits, which enable pictures to be *intercut, inter-mixed/dissolved,* combined (as *superimpositions,* or one picture *inset* into another), or juxtaposed (*split-screen*).

Two or three cameras are sufficient for general purposes, but for large productions four or five cameras (sometimes more) may be used to allow continuous shooting. In a few studios, instead of interswitching during the performance, each camera has its own associated video recorder (*dedicated VTR*), their separate video recordings being combined later in a *post-production editing session.*

Teamwork

There are several important differences, for a cameraman, between shooting a production with his single camera and being a member of a camera team. The *single cameraman* as part of a production group often selects the optimum shots for himself (as in ENG, and other location assignments), particularly if he is doing the jobs of both director and cameraman.

A cameraman within a *multi-camera crew,* on the other hand, is working to instructions over the *intercom (talkback)* system from the director in the *production control room.* He needs to coordinate his efforts with others, so that the pictures match and can be interrelated to form a smooth-flowing presentation.

Continous shooting

The advantage of using several cameras during continuous production is that the director is able to introduce *instant* visual changes. He can intercut between different sizes of shots, follow action as it moves from one area to the next, combine shots in various ways and vary the audience's viewpoint.

Perhaps more important is the fact that this method of shooting is capable of providing a *complete production-package in a single session,* either live or on tape. When taping, of course, one has the option of re-recording and editing-in, to correct any faulty passages. (Single-camera shooting usually results in a series of separate 'takes' that have to be *edited together later.*)

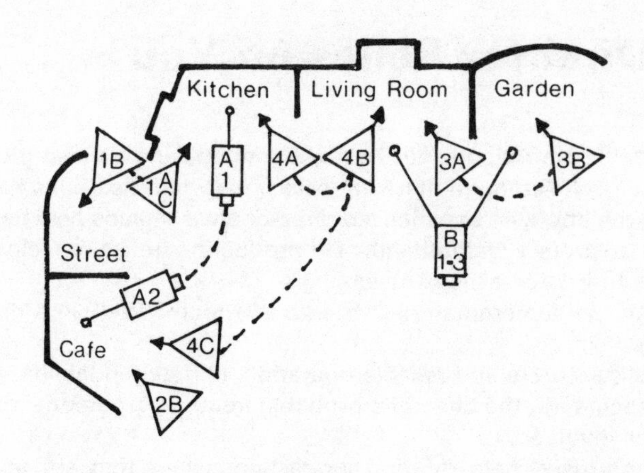

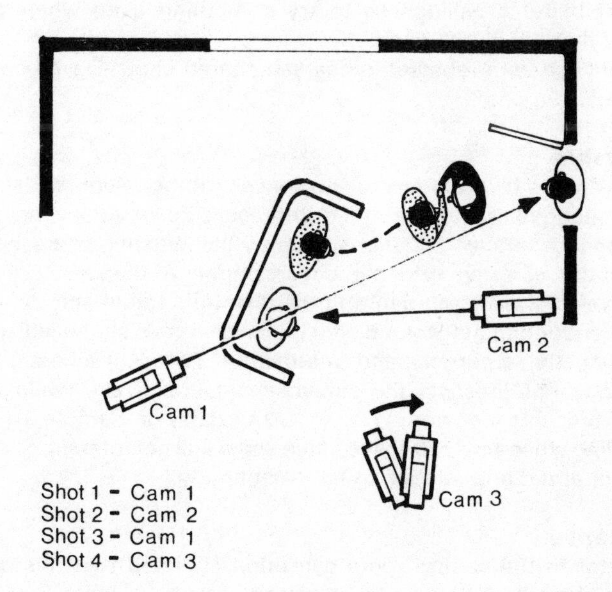

The camera plan
The positions of cameras 1,2,3,4 are successively marked (1A,1B), and positions of sound booms A, B are shown as A1, A2.

Multi-camera shooting
Each camera is allocated its shots: e.g. Shot 1 – Cam 1; Shot 2 – Cam 2; Shot 3 – Cam 1; Shot 4 – Cam 3.

Get a clear idea of what the director is aiming to do. If you are unsure, ask.

The Director Relies on You

'On-the-spot' inspiration is not enough when making a video program of any complexity. Although there are occasions when decisions have to be made 'on the hoof', an experienced director always plans how he is going to tackle a project beforehand. TV production relies on closely knit teamwork to interpret those ideas.

How can the cameraman best fit into this team? Well, in a number of ways:

1. By skill, accurate and reliable operation, and dependability.
2. By anticipating the director's probable treatment, potential problems, mechanics involved.
3. By unobtrusive help, offering suggestions where they are needed and are appropriate to the production.
4. By flexibility, a willingness to try something even when you are convinced that it will not work!
5. By patience with repeated rehearsal, altered shots – even confusing instructions.

Be adaptable

There will always be problems of one kind or another. Perhaps there is not time for a planned move, or a shot is impracticable, or a dolly move needs to be changed. In such situations, using your initiative or suggesting a viable solution can save valuable rehearsal time. At the same time, guard against over-zealously reorganizing the director's treatment!

It is not enough to get 'a good shot'. Each camera's shot needs to relate to the production's purpose and treatment. If Tom on Camera 1 decides that a high shot presents the subject most effectively, while Dick on Camera 2 presents a powerful low shot, and Harry on Camera 3 goes into screen-filling close-ups, the consecutive shots will not intercut. Shots have to be rationalized and related, to have continuity.

Use initiative

The director in the control room can often only see what his monitors show. So during rehearsal a cameraman may help, when off shot, by giving a wide view which may reveal unsuspected studio problems – e.g. that a sound boom is blocking a camera from moving in closer. In unrehearsed, impromptu or unpredictable situations, a cameraman might offer up unanticipated shots from his viewpoint. Remember, though, that unplanned shots may frustrate other members of the production team.

Be consistent

Always try to repeat shots exactly as rehearsed. Only modify your shot if circumstances make this necessary, e.g. if a performer is out of position.

Confirming the shot
Make sure you are getting the required shot. These are all *two-shots* but each has quite a different audience impact.

Correcting shots
When errors arise, do what you can to compensate. In this case the right-hand person has not walked forward sufficiently to hit his floor marks. The camera moves round to improve the shot.

Ready for Rehearsal

Here is a brief run-down of the various routine pre-rehearsal checks a wise cameraman makes. Most seem obvious . . . but that is when you start to take things for granted!

Camera check-out

1. *Preliminaries.* Camera switched on, warmed up and lined up?
2. *Camera cable.* Are the cable plugs at the camera and wall (or equipment) socket tight? Is the cable secured to the camera mounting? Is there sufficient cable for camera moves and is it suitably routed?
3. *Camera head.* Is the pan-bar/panning handle(s) firmly attached and at a comfortable angle? Unlock the tilt secure control and check the camera-head balance. Is it nose or back heavy, or needing readjustment? Check (and adjust) vertical drag/tilt friction. Unlock pan-secure. Is the action smooth, with just sufficient friction? (Adjust horizontal drag/pan friction).
4. *Column.* For a rolling-tripod, check column height and adjustment. For a pedestal, unlock column and raise/lower it. Check the ease of vertical movement (balance) and adjust if necessary.
5. *Steering.* Check freedom of movement in all directions, in dolly (track) and truck (crab) modes.
6. *Cable guards.* Adjust guards to prevent cable overrun or floor scraping.
7. *Lens.* Remove the lens cap. Switch lens electrical capping out. Check that the lens is clean.
8. *Viewfinder.* Check its focus, brightness, contrast, picture shape (aspect ratio), edge cut-off, image-sharpening (crispening). Are internal indicators working (for zoom, *f*-stop, exposure, etc.)? Are tally and indicator cue-lamps OK? Check mixed viewfinder feeds.
9. *Focus.* Check focus-control smoothness from nearest to furthest distance (infinity). In a lens servo system, is there any spurious focus overrun or hunting (rhythmical changes) occuring? Is the zoom action smooth throughout? Focus at the widest and narrowest angles. Check that focus is constant throughout the zoom range (tracking correctly).
10. *Lens aperture.* Check the *f*-stop number selected during camera line-up with chart under standard illumination. (This indicates system's sensitivity.)
11. *Zoom.* Check zoom meter or indicator. Adjust shot-box, if any, to preset angles. Check operation.
12. *Intercom (talkback).* Check intercom circuits (general and private-wire). Is program sound feed OK?
13. *Filters.* Check filters (filter wheel), making sure that one is not accidentally left in position.

Preparing portable cameras

In addition to the above, check out the following:

1. *Cable.* Do you have enough cable for the job? Has the camera cable-compensation control been set to suit the cable length? Are cable joints firmly screwed together? Route the cable to protect it from feet, vehicles, etc. (sling, hide, cover over, bury).
2. *Battery.* Check that the fitted battery and standby batteries are fully charged.
3. *Viewfinder.* Is the fitting secure? Check warning lights, including tally-light, exposure, low battery, VTR on, tape left, etc.
4. *Microphone.* Is the mike secure and functioning? Check connections. Cable for announcer/commentator mike.

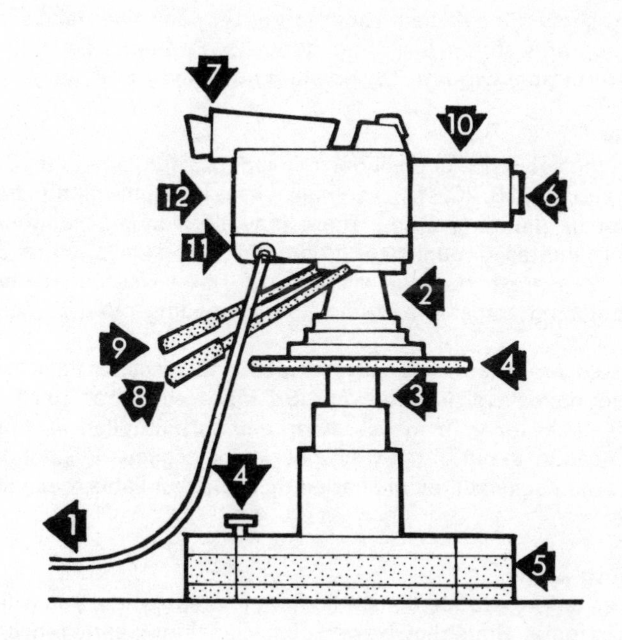

Camera checkout
Typical checkout points include:
(1) Camera cable. (2) Camera head features. (3) Column. (4) Steering. (5) Cable guard. (6) Lens condition. (7) Viewfinder. (8) Zoom action. (9) Focus checks. (10) Lens aperture. (11) Intercom (talkback). (12) Shot card (camera card).

Check around to familiarize yourself with the scene of action.

Check the Studio

Begin by identifying the various action areas. It may be obvious: here is the laboratory, there is the office set and beyond is the interview spot. But suppose there are several interview or performance areas? Which is which? Often they are given identifying letters or names.

Using your *camera card,* locate the various basic camera positions and moves. Are there any obvious problems? Perhaps there is some obstruction just where the dolly will be operating. It happens in the best-organized shows; insufficient room to get between two flats, some properties temporarily stored, a lighting fitting . . . . Perhaps a stand-lamp is just visible through a window. Observation now saves time later.

Cable routing

Remembering the sequence of positions marked for your camera (e.g. for Camera 2 marked: 2A, 2B, 2C, etc.), estimate where its cable will run, from the wall outlet or plugging point. There may be a single communal wall-box for all cameras, or duplicates around the studio walls so that you can use the one nearest the main action area. This saves unnecessary cable runs stretching across the studio floor, impeding other cameras' movements.

Camera cables are relatively easily trapped under scenic flats, floor monitors or loudspeakers, furniture etc. So make sure that you have enough active cable for your moves. Keep any surplus piled in a neat figure-of-eight coil in an out-of-the-way spot behind scenery. It is not only frustrating to strain against tight or snarled-up cable, but liable to damage the cable too.

Check dollying areas

It is a good idea to look around the studio floor in areas where you will be dollying your camera. There may be odd obstacles during early rehearsal – ranging from cables (lighting, monitors), lumber used in last-minute construction, carpets, even wet floor paint – that are going to bug smooth dolly moves. The floor needs to be clean, too. A dropped cigarette butt has stuck to a dolly wheel before now, and caused camera judder!

Ready to go

Finally, let's move to the opening camera position. On with the headset, just look over the camera cards (shot-sheet) or run-down sheet, focus up on the scene and you're ready to go!

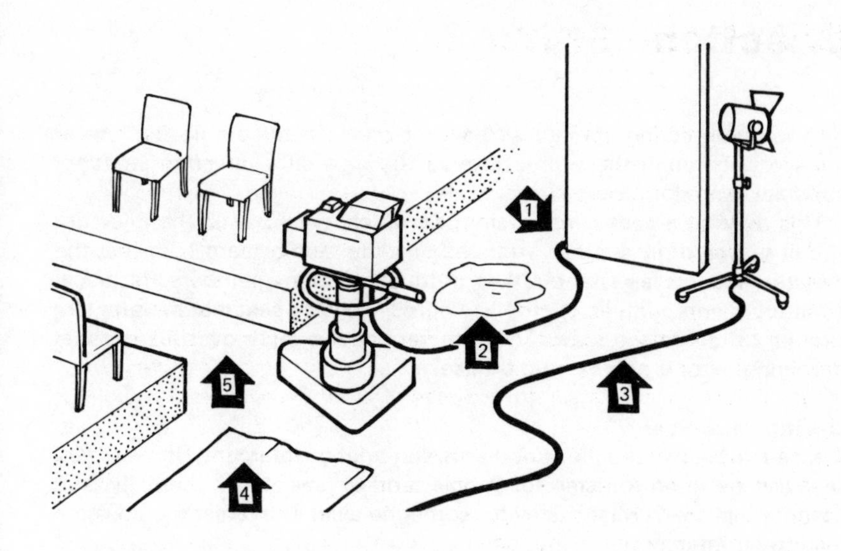

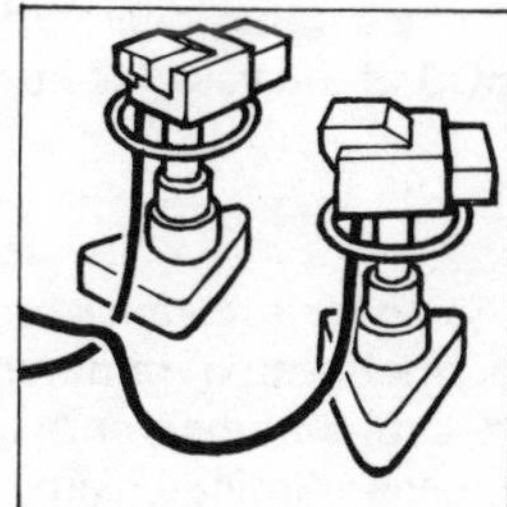

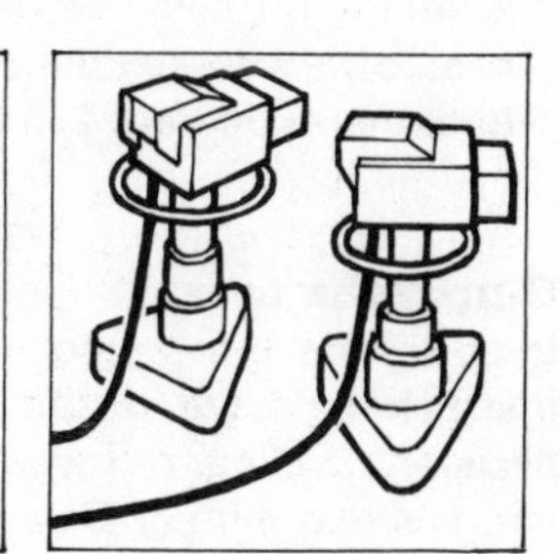

Floor check
There are often various problems for the unwary, including:
(1) Camera cable trapped. (2) Water on the floor. (3) Lighting cables that prevent dollying. (4) Floor covering that frustrates a dolly move. (5) Scenery that prevents camera getting into position.

Cable routing
In a continuous multi-camera production, it can be important to arrange camera cables carefully so that they do not prevent dollying. Typical arrangements include: working in another camera's loop, working under another cable (best avoided) and parallel routing.

Unheard by the talent, the intercom coordinates the production crew.

In Action

All members of the camera and sound crew on the studio floor wear headsets. Through these they receive the general *production intercom (talkback)* and the program audio.

This may be a communal system, in which everyone on the intercom circuit is heard. In another arrangement, the studio team hear just the director and his assistant. Other production-team members (technical director, vision engineers, etc.) use access-keys or separate *private-wire* circuits to talk to the crew. The cameraman can reply over his headset microphone, or a nearby studio mike.

During rehearsal

It is best to keep instructions or discussion brief on intercom. Unnecessary chat can be tiring to listen to. People tend to 'switch off' mentally, and assume that the verbiage is all for someone else! The director guides the team over intercom:

> 'Camera 1, he will be getting up in a moment, so start to pull out (the seated person stands) . . . he's going over to the table. Tighten as he gets there. Camera 2, a BCU of the vase, as he points to the decoration . . .'

During the take

By the time the taping session or transmission time arrives, intercom instructions have become brief action reminders, with the director's assistant calling out each shot number and camera, readying other sources (e.g. film channel), giving timings, countdown cues, etc.:

> 'Coming to 1 . . . On 1 . . . Shot 15 . . . Ready for the rise . . . Stand-by 2 . . . Coming to 2 . . . On 2 . . . 1 move to the window . . .'

From his camera card, the cameraman knows details of his position, type of shot, etc. From rehearsal, he has learned the performer action, camera operations involved (panning, focusing, dollying . . .). Now he only needs a brief recall. So intercom coordinates action, reminding the crew of any particular difficulties.

Be prepared

Always keep a step ahead by checking on your next planned floor position and its shot(s). When your present shot is completed, move there as soon as possible: silently, yet quickly.

When cameras move fast, or over long distances, the noise of cable drag can be heard quite clearly during quiet action. A cable-handler can assist here (*cable grip*) by ensuring that the camera always works with a sufficiently slack cable loop, collecting together surplus cable, avoiding snarl-ups, taut or trapped cable and similar hazards.

Instructions		Meaning
Stand by 2: ready 2: coming to 2 . . . on 2, shot 40, camera 1 next		General guide procedure for cameras (used for general crew co-ordination).
Pan left (or right)	head movement	Turn the camera head left (or right).
Tilt up (or down)	head movement	Tilt the camera head up (or down). 'Pan' often used, e.g. 'pan up'.
Centre up (frame up)	head movement	Arrange subject in centre of picture.
Focus up	focus	Focus hard on subject (to a defocused camera).
Lose focus on . . .	focus	Defocus subject named.
Split focus	focus	Focus evenly between subjects (making both of optimum sharpness).
Follow focus on . . .	focus	Keep hard focus on moving subject named.
Focus forward (or back)	focus	Refocus nearer (or further away).
Defocus the shot	focus	Defocus shot overall (usually by focusing forward to limit).
Give more (or less) headroom	framing	Increase (decrease) space between head and upper frame.
Cut (or frame) at . . .	framing	Compose picture to place frame at subject named.
Lose the . . .	framing	Adjust framing to omit subject named
Stand by for a 'rise'	framing	Be prepared for performer to stand up.
Single shot	shot type	Shot containing one person.
2-shot, 3-shot, etc.	shot type	Shot containing number of persons indicated.
Group shot	shot type	Shot containing an indicated group of persons.
Close-up, mid-shot, three-quarter shot, etc.	shot type	Specific proportions of people filling frame.
Wide shot (cover shot, long shot)	shot type	General term for an overall view of action from that camera position.
Give me a wider shot	shot type	Use wider angle lens from that camera position.
Widen the shot a little	shot type	Usually indicates slight track back to increase coverage. May mean increasing zoom lens angle.
Give me a looser (fuller) shot; not so tight	shot type	Provide more space in frame around subject(s), i.e. longer shot.
Tighten the shot	shot type	Reduce space between subject and frame, i.e. closer shot.
Track in (dolly in)	dolly movement	Move camera mounting towards subject.
Track back (dolly back)	dolly movement	Move camera mounting away from subject.
Creep in (or back)	dolly movement	Move camera very slowly towards (from) subject.
Crane left/right (also tongue, slew, or jib)	dolly movement	On a crane, swing the camera boom arm to left (right).
Crane up/down (also boom or tongue)	dolly movement	On a crane, swing the camera boom arm up (down).
Elevate (ped up)	dolly movement	On pedestal mounting, raise camera head (i.e. alter lens height).
Ped down (depress)	dolly movement	On pedestal mounting, depress camera head (i.e. alter lens height).
Tongue in (or back)	dolly movement	With crane base at right angles to subject, swing bottom arm to (from) it.
Zoom in (or out)		Narrow lens angle (widen lens angle).
Clear on three		Camera 3's shot is now finished. Move to next position. (Dismiss subject.)
Clear two's shot		Remove obstructions from camera 2's shot.

Getting a good shot is one thing; repeating it to order is quite another.

Lining Up Your Shots

A skilled cameraman not only gets effective shots during rehearsal but can repeat them on air, despite the tensions and complications of transmission conditions.

In a static (stationary) arrangement, where someone is sitting behind a desk, it is easy enough to repeat the same shot. But in a dynamic situation where people move around, groups form and reform, accurate composition under changing conditions can be quite critical.

Repeated shot accuracy

You can use a number of techniques to ensure that you get the same shot accurately each time:

1. *Check your lens angle.* No problem of course, with a fixed-lens system. But because a zoom lens's angle is variable, you need to make sure that it corresponds with the angle used during rehearsal. If it doesn't, proportions will be different.
2. *Check 'landmarks'.* Look at a shot carefully to see where the frame 'cuts' objects in the background. These clues help you to remember the original composition.
3. *Check surrounding space.* What is nearly coming into shot? There may be a bright window to avoid. Are you nearly off the set (*overshooting/shooting-off*)? By anticipating shot limits, you will not be caught out.
4. *Check the dolly's floor position.* A quickly made *floor mark* can show exactly where the camera was located in rehearsal. Use only adhesive tape, crayon or chalk marker (never permanent pen) so that the mark can be removed later. Avoid conspicuous or excessive marks, for they can be confusing and may be visible if that area of the floor appears in other shots. You can also use parts of the setting or furniture to assist you in locating a camera position accurately, e.g. the end of a wall or beside a bookcase.

Why bother?

Some cameramen rely solely on their 'pictorial memory' and spurn such aids. They are the ones who often encounter unexpected lens-flares, camera shadows or compositional errors, for their guesses can prove less accurate than they believe. You can easily find yourself working at a different distance from rehearsal, and using another lens-angle to try to compensate. You may dolly over to a place that seems 'near enough', but find that from this spot the background features look noticeably different behind the subject. It is not surprising that so many experienced cameramen insist on methodical shot line-ups.

Floor marks
Mark your basic floor positions carefully, always using the same part of the dolly as a guide.
In a multi-position show, lettered marks can relate to your camera card. Avoid using too few or too many marks. This may only confuse you.

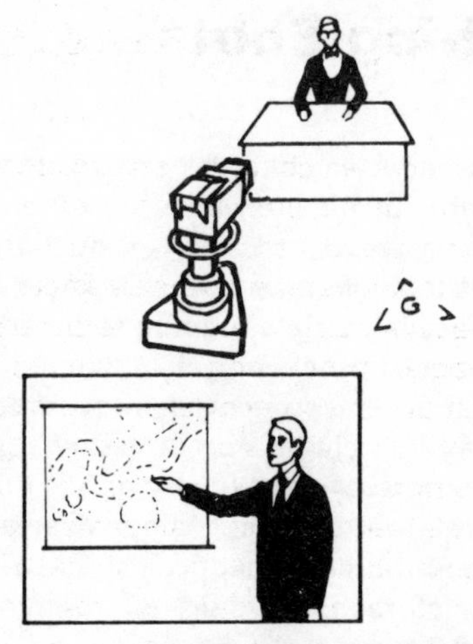

Shot coverage
Check your lens angle. Look around to check what is just in shot and what remains just outside the frame. By doing this you can repeat the rehearsed shots exactly during 'transmission'.
If the shot omits items that were originally included, or now shows extra items, its impact changes.

Nearby subjects
Keep an eye on what is happening in vicinity. People may move into shot accidentally. You may find that just a slight camera movement brings unwanted things into shot.

Remember, individual shots have to be cut together.

Matching Shots

As an *audience* watches the screen, they see a sequence of carefully selected intercut pictures telling a story. But the *cameraman* (whether alone or in a crew) sees any production as a series of isolated, often unrelated shots. He may not even know how his particular parts of this complex jigsaw puzzle are going to be edited into the complete show.

A well-made production has continuity of styles and techniques. Although shots have been taken on different occasions or by different people, they intercut as a smooth-flowing continuous effect.

So the cameraman has to rely on the director's guidance to ensure that each shot relates to the next. Otherwise, left to themselves, a camera team might choose almost identical shots of the same subject or provide unrelated pictures that would not intercut successfully.

Avoiding spurious effects

If several cameras are taking shots of similar subjects, it is possible to get some very strange effects when cutting between them. Different-sized pictures of the same subject can make it appear to grow or shrink instantaneously on a cut. Cutting between similar shots of two people can produce a startling 'Jekyll and Hyde' transformation. When the headroom differs in two successive shots, or camera heights vary, the visual jump can be quite distracting on a cut.

Although these are mainly matters for the director to coordinate when arranging shots, it is important for the cameraman himself to understand the problem and to relate his shots as far as possible.

Matched shots

There are situations where you want to match pictures from different cameras so that their shots combine very exactly – for example, where one camera is shooting a graph grid, while a second shows data that is gradually uncovered. Such precisely registered effects can be quite tricky owing to picture-geometry variations between cameras and the time needed to align the shots.

Mixed feeds

Many video cameras have a special switching facility (*mixed feeds*), allowing the cameraman to superimpose another selected camera's shot on his own viewfinder picture. He can then adjust his camera to make the two fit. Otherwise he has to follow a composite picture on a nearby monitor or be guided into position from the production control room.

1 2

3

4

Accidental effects
As the director cuts from one picture to another, visual mismatching can create disturbing effects.

1. *Height jumps.* Caused by mismatched lens heights that have no dramatic purpose.

2. *Headroom changes.* Shots have different amounts of headroom.

3. *Size jumps.* If the subject changes size slightly on a cut, it creates the effect of instant growth or shrinkage.

4. *Transformations.* If shots are very closely matched, one person can seem to become transformed to another on the cut!

Properly planned, the audience need never realize that we are making it up as we go along!

Shooting Unrehearsed Action

Rehearsals are important! It is the director's opportunity to judge whether his ideas and organization are going to work. He wants to check action, performance, shots, sound, lighting, staging and the various other contributory elements as fully as possible.

But some situations cannot be rehearsed at all because they are once-only events, or are virtually unpredictable. Instead of positioning cameras and simply hoping for good shots on-the-fly, an experienced director devises a plan of campaign, a flexible treatment that will give him the best options and can be adapted as circumstances allow.

Preparation

Some types of program can be so familiar to a camera team (e.g. interviews, game shows) that they only need briefing on their own individual contributions.

A director may outline the nature or purpose of the show, to help his crew in selecting shots. Where, for example, a group of speakers is going to be questioned by members of a studio audience, the operational problems vary according to whether people are going to speak in a prearranged order or talk at random.

It is also helpful if the team is given the names of the speakers or characters in a production. It is embarrassing to be asked for a close-up of John, only to find that you've picked Bob by mistake!

Methods of approach

There are two ways of tackling unrehearsed action:

1. By allocating certain shots, e.g. Camera 1 takes *wide shots* (*cover shots*) of the area while Camera 2 concentrates on *closer shots* as they become available.
2. The director, who is isolated in the control room, may largely rely on the initiative of each cameraman to provide him with optimum shots. He can select from their coverage although he cannot, of course, judge what is happening out-of-shot.

'Grab shots' are fine, as long as a cameraman does not just concentrate on the items that interest him. A series of shots of pretty girls in the crowd may not really relate to the show's purpose! You also have to guard against all cameras concentrating on the same aspect of the scene and completely overlooking others.

The best method of selecting grab shots is to *prefocus the zoom lens* on a close shot, then zoom out to a wide view. Keeping your head away from the viewfinder, check around for appropriate action. At any sign of a useful subject, zoom in to detail for the director to assess.

Area coverage
Cameras may concentrate on certain areas, ready for action there.

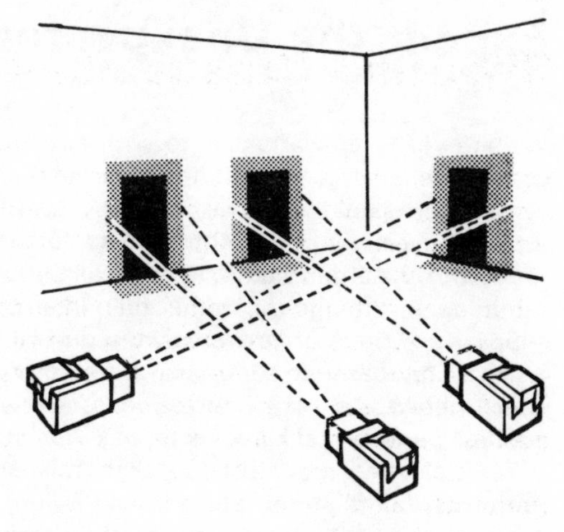

Localized coverage
Cameras may concentrate on certain types of shot, moving to fresh areas as directed.

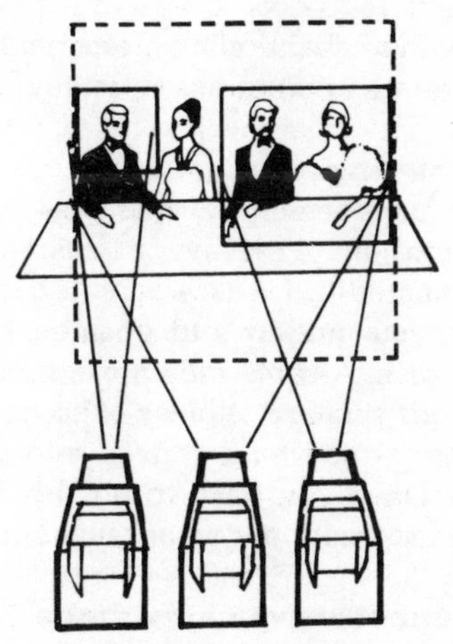

Unexpected action
By checking general action outside your shot, you can see if you are missing any important action.

Expect the Unexpected

We are really concerned here with two things: *safety* (i.e. preserving oneself, other people, and the equipment) and *keeping the show going.*

Many productions are a once-only event, where even one person's inattention can mar the entire team's efforts. Various things can go wrong – and do! But if problems do not affect you directly, it is better to leave others to sort things out, rather than interrupt your own contribution.

Some accidents are extremely disconcerting: an object falls off a table; a lamp explodes with a loud bang; a stepladder falls to the ground. Other catastrophes are less impressive: a cable pulls a floor-lamp out of position; a water tank begins to leak . . . . But they can all be distracting.

Panic can become infectious, slight mishaps can become disasters. If a performer faints during a live transmission, it could be more helpful for the camera to dolly over a little to make room and carry on with his shot, rather than abandoning his camera and going to help. Other people are in a better position to assist without disrupting the show.

Improvising

Certain unforeseen events may affect you but not your function (as a cameraman receiving a custard-pie during a knockabout routine appreciated!). If it is a continuous or a live production, and something stops you getting into position for your next shot, do your best and improvise. A cable may prevent your dollying in; water may have made the floor slippery; a piece of scenery may be out of position. On location, someone may have moved into your path. Take it philosophically and press on. If you have to provide a substitute shot, make it as near the intended treatment as possible.

The unobtrusive catastrophe

Most cameramen have their own stories about why a camera was late on a move (it had stuck to the floor-paint); or why a camera did not move to the next shot (he was holding up some scenery about to fall on him). The greatest compliment any director can pay his crew is to have been completely unaware that anything was wrong!

The helpful warning

Although a busy cameraman mainly concentrates on his own activities, he can sometimes help others by anticipating trouble – warning them over intercom, for example, that a keylight has failed or a chair is out of position.

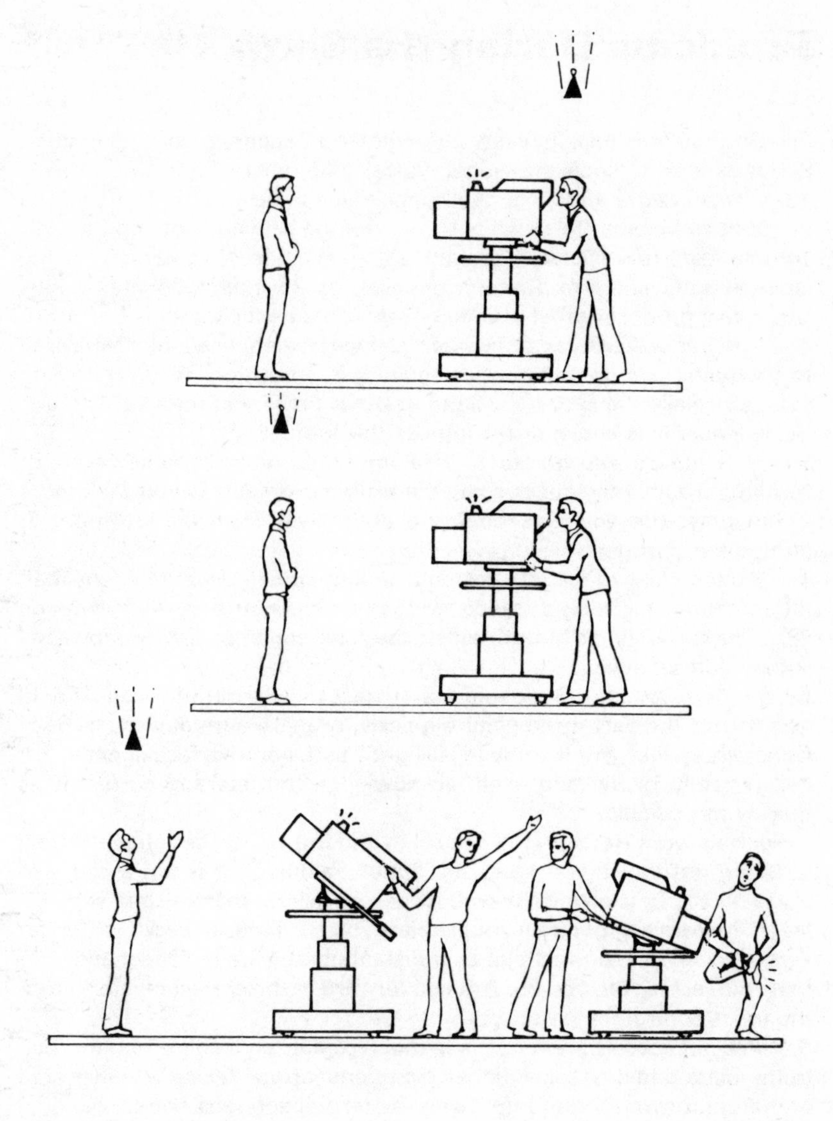

Be alert
Some disasters directly concern the cameraman himself . . . and you should take appropriate action.

Concentrate
Do not let other people's problems distract you – although it is friendly to alert them.

Confusion can spread
One person's problems can become another's . . . and disrupt the program.

Problems During the Show

Certain problems arise in every cameraman's experience sooner or later. So let us look at these, and some typical solutions.
1. *'I've missed a shot!'* It can happen to the best . . . but rarely. The moment comes for the switcher to cut, but the camera is not on its shot! The situation is universally embarrassing. Whatever the reason for the lapse, if sufficient shot duration remains, get to it as soon as possible. Otherwise make sure that you do not miss the next shot too.
2. *'The performer's out of position!'* Even experienced talent may fail to hit their previous marks when moving into position. When recording you can retake the faulty shot, but at other times you may have to arc, truck, adjust lens angle, or recompose the shot.
3. *'He's moved into my shot!'* Having set up your shot, someone or something suddenly intrudes into the picture. Sometimes you can ignore it, but otherwise you can reframe slightly away from the intrusion, or tighten the shot gently.
4. *'It's too close to focus!'* When a subject comes closer to a lens than the minimum focusing distance, so that you cannot focus sharply (page 28), you can only dolly out, widen the lens angle or simply move the subject further away.
5. *'I've lost focus!'* Suddenly the subject moves out of focus. should you correct the defocused picture quickly, or make a gradual correction? Generally speaking, if the image is slightly soft, improve focus gently. But if it is really badly defocused, acknowledge the fact and correct it as quickly as possible.

Rocking focus (rotating the control to and fro) will show you the correct focusing position if you have any doubts, although it is not a check to make on air! To see which way to focus, check how sharp distant subjects are. If they are sharper than your subject you are 'focused back', i.e. too far away (beyond the correct plane). If distant objects are indistinct and your main subject is soft, you are focused forward of the correct plane, so turn the focus control the other way.
6. *'I've got a lens flare!'* Usually much clearer on a color monitor than in the black-and-white viewfinder. Solutions include taking a higher shot and tilting down a little; improving the lens shade/lens hood; shielding the light off the camera.
7. *'My cable is trapped!'* There is little you can do except move around within the available cable length until there is an off-shot opportunity to clear the obstruction. Otherwise you have to vary the lens angle to substitute for dolly moves.

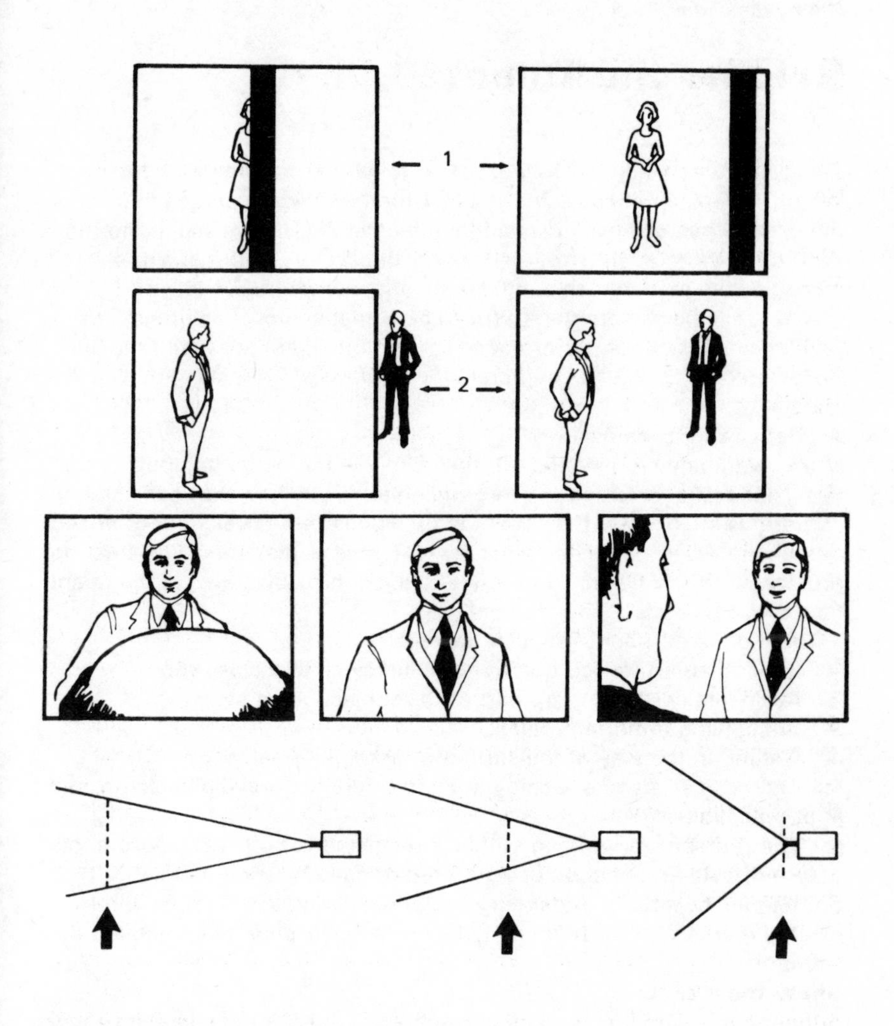

Wrong positions
A performer moves into the wrong position and (1) becomes masked by scenery (or another performer), or (2) is left outside the shot. By trucking (crabbing) left, the dilemma is resolved.

Sudden intrusion into the shot
If someone near the camera accidentally moves into shot, you can reframe the shot or move the camera to avoid him. Alternatively you can make him a definite part of the composition.

Minimum focusing distance
If a subject comes closer than the lens's minimum focusing distance, you have to pull back, zoom out, move the subject further away . . . or let them go out of focus.

Helping the Talent

The cameraman can do a great deal to put a performer at his ease. Whether an experienced actor or a first-time visitor, the person in front of the camera has a sense of isolation, a feeling that now it's all up to him. Without taking on the role of second director or bypassing the floor manager, the cameraman can help in various unobtrusive ways.

Above all, the cameraman needs to be something of a diplomat. Even a regular professional's performance can fall off when faced with the 'cold turkey' reception of an indifferent floor crew. A friendly grin can be infectious.

Advising the newcomer

Many who appear in front of the camera are quite unfamiliar with television techniques and are very willing to be told how they can help the cameraman to get the best shots possible. Even experienced demonstrators or interviewers may not realize that they are creating problems. This is where the cameraman can help by showing the talent how they can assist.

There are a number of regular hazards:

1. Moving an article too quickly to be followed in a close shot.
2. Accidentally covering up part of the subject being shown.
3. Shadowing important detail.
4. Getting in the way of the subject (masking).
5. The need to handle a shiny subject carefully, to avoid catching bad light-reflections.
6. The problem of working within a restricted area in very close shots (depth-of-field limitations), using a pre-arranged location mark.

Sometimes by sitting or standing in a certain way, or by hitting carefully located *marks,* a performer can help with a tricky shot if you explain the situation.

Show the director

Although it is sometimes more convenient to direct a multi-camera show by standing near the camera and using a nearby picture-monitor to rehearse shots, most directors prefer to sit in the production control room in this situation and assess operations from the preview monitors there (one per camera channel).

Studio problems can sometimes arise which are all too obvious on the spot, but quite unsuspected by the director in the production control room. This is where a helpful wide shot on a nearby camera can reveal the situation, showing, for instance, that the talent is at the edge of a platform, so cannot move the couple of paces the director requires.

Remember the prompter

When performers are reading their lines from a prompter attached to the camera, you can often make life easier for them by ensuring that your camera is within comfortable reading distance, and not so high that it forces them to look up into the lights while reading.

Putting them at ease
A kind word or a little guidance can put inexperienced talent at their ease.

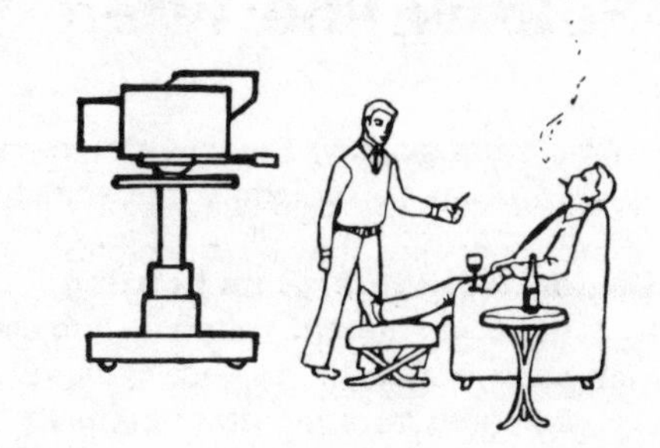

Reveal the problems
Performers may have difficulties that are not obvious to the director in the production control room.

Show them the shot limits
Let the talent know if you want him to work within a very restricted area. This is best demonstrated on a nearby picture monitor.

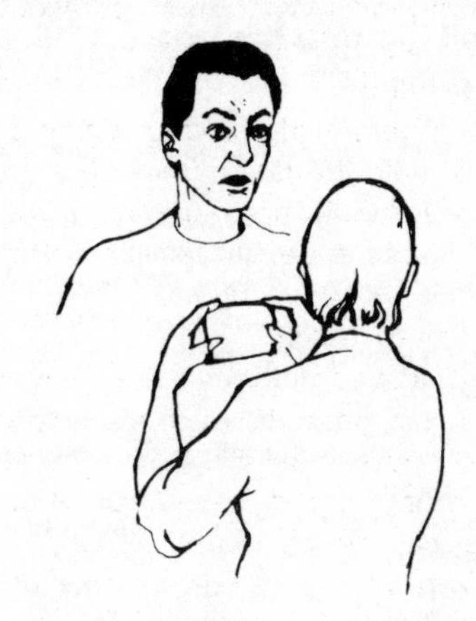

Remember the prompter
Ensure that you are sufficiently near for the performer to read a camera-attached prompter.

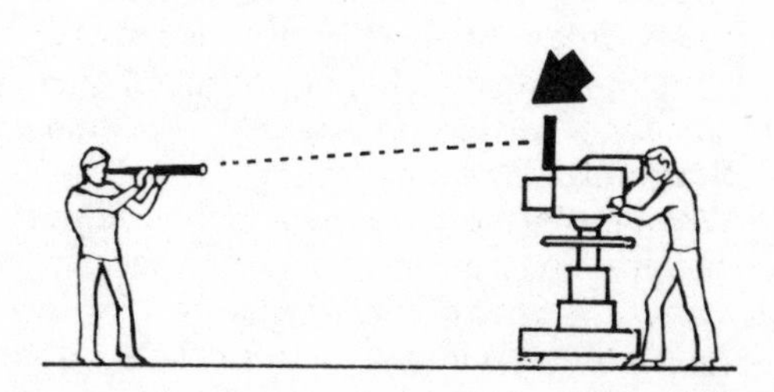

Helping Lighting Treatment

Your picture displays the result of many individual skills. Exactly how you shoot a scene can strongly influence the impact of this work.

Camerawork interprets lighting

If you decide to arc round a subject to what you consider a better viewpoint, and 'sell' this shot to the director, are you showing initiative or are you re-interpreting other people's work?

The *lighting director* may not welcome the change, for his carefully angled backlight, intended to produce an attractive rimming from the *planned* viewpoint, is now behaving as a harsh frontal keylight instead! The *set designer,* who was not expecting the scene to be shot from this angle, now has to modify and re-dress the set to suit the new situation. The *audio man* may now have problems with boom shadows when trying to pick up sound for this unplanned shot.

The effect any light has on the appearance of a subject alters with the camera's viewpoint, and this is taken into account when the lighting treatment is arranged. Remember, it is a lot easier for the cameraman to move his camera, or for the talent to be given new positions, than it is to change the lighting and scenery to suit new, unanticipated camera angles. Alterations take time and may upset other shots.

Principal lighting hazards

Let us summarize typical lighting considerations that affect the work of the cameraman:

1. *Camera shadows.* The camera's shadow falling onto the subject or being seen in shot.
2. *Lens flares.* Often not obvious in the monochrome picture, but usually cleared by adjusting camera height, extending the lens shade, or barndooring the lamp.
3. *Spurious light reflections.* Direct lamp reflections on a shiny surface (e.g. glass windows), or reflections of light on glossy areas, can often be avoided by slightly altering the camera height or position. (Otherwise the lighting or surface-angle may need to be changed.)
4. *Shading in graphics.* When shooting graphics or title cards that are lit from above, you can usually adjust your camera height to avoid uneven illumination (bottom shading).
5. *Changing effective lighting angles.* Keylight positions are chosen so that they are roughly 10° – 40° from the direction a person is going to face. If the camera trucks or arcs and the talent looks at the camera, their angle to the keylight changes and this coarsens or flattens the lighting.
6. *Lights appearing in shot.* Cameras can easily shoot off into backlights behind the subject, particularly from low viewpoints. Where lamps or specular reflections appear in a scene, take care that they do not burn onto the camera tube, leaving permanently defacing marks.

Lighting matches planned shots
If instead of the planned position, 1, the cameraman moves to position 2 to improve the shot, he can upset lighting treatment. Here the original backlight has now become a frontal keylight.

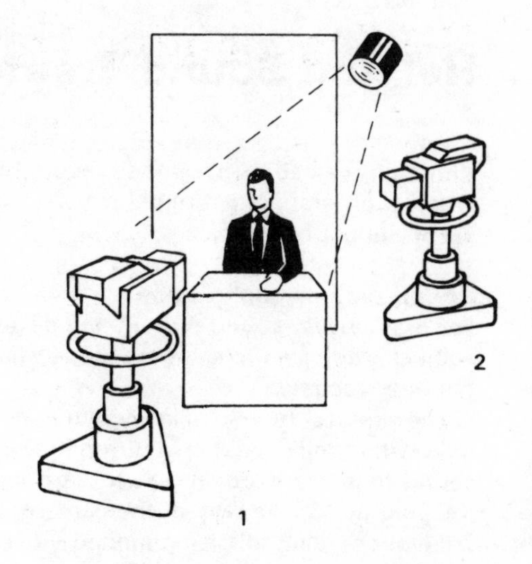

Camera shadows
If a close camera casts shadows on the subject, a more distant position with a narrower lens angle can clear the problem.

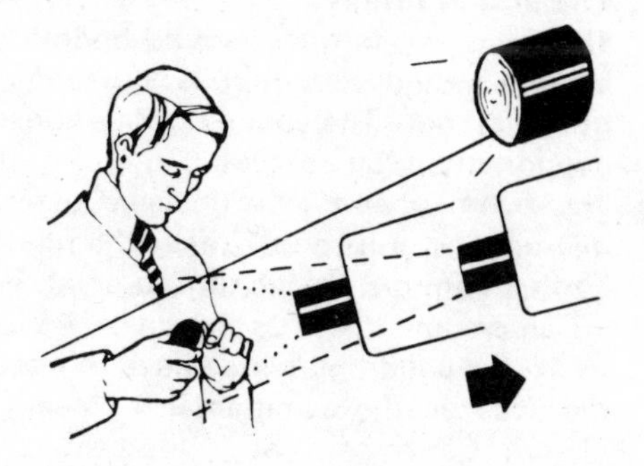

Lens flare problems
Lens shades (sunshades, lens hoods) for zoom lenses are most effective at the widest angles. At narrower angles, lens flares sometimes arise which can be cured by a temporary strip attachment.

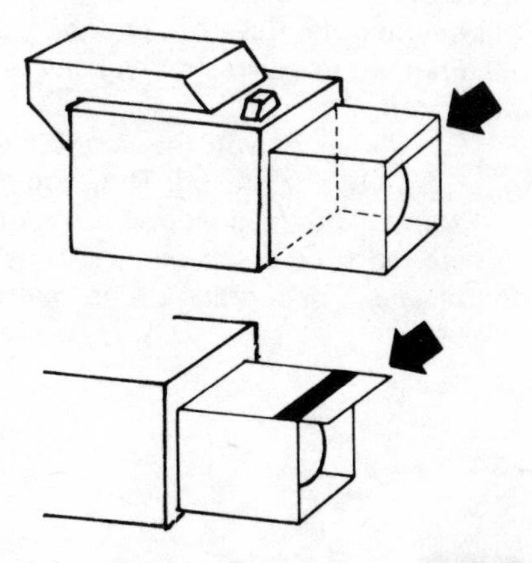

Without the cameraman's cooperation, efficient boom operation could be impossible.

Helping Sound Treatment

Although we see microphones regularly on the screen, there are many types of program (e.g. drama), where a visible mike would be distracting and quite out of place.

Sound pick-up techniques

For high-quality sound pick-up, the distance of the microphone from the subject is adjusted to suit the *shot size* (i.e. match sound perspective) and the *local acoustics.*

The simplest mobile mike support is the *fishpole* (*fishing-rod*). This is a pole with a mike fixed in a stirrup at the far end, held out of shot by the sound man – low down below the frame of the shot, or high above it. Keeping out of the way of the camera, he does his best to position the mike out of shot, following limited movement and avoiding shadows.

The sound boom

There are two forms of sound boom: the *small boom* (*giraffe*) used for stationary and semi-mobile situations, and the *large boom* used for general action. The sound boom's horizontal arm is counterbalanced and supported on the center-column of a wheeled stand or mobile platform. The boom operator swings and extends the boom arm to place the microphone at its end, exactly the right distance from the sound source. Further controls tilt and turn to point the mike as required.

Camera and boom operations need to be coordinated. When a voice is weak, the sound man may have to place the microphone much closer to the speaker, the cameraman tightening his shot to keep it out of the picture. Conversely, when the cameraman needs more 'air around the subject' (wider shot) than usual, the boom operator may have to work with his mike further away to enable the cameraman to get his shot.

To summarize, then, the cameraman can assist sound men in various ways by:

1. Keeping the mike out of shot (careful framing; adjusting headroom).
2. Framing to avoid showing any mike shadows cast on the scene or the performer.
3. Coordinating with the sound-crew as they move around the studio floor. Space is restricted. They could easily get in each other's way!

Although boom shadows are largely a matter between the lighting director and the boom man, a cameraman can often avoid or aggravate the problem, depending on his exact dolly position and framing.

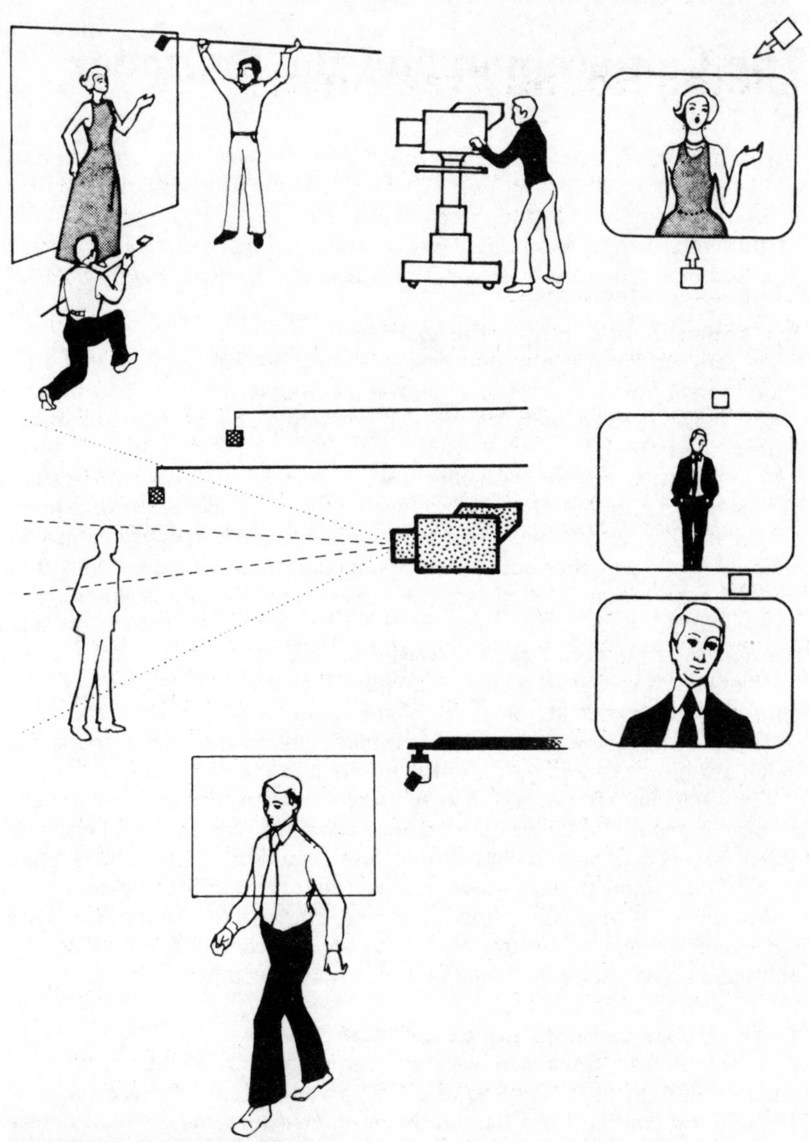

Keeping the mike out of shot
The sound man aims to hold the microphone as near to the performer as necessary without moving into the picture or casting a shadow. The cameraman may frame the shot to help him.

Matching sound and picture
The sound man positions his mike to suit the shot size. Close shots need close mike positions. Longer shots need a correspondingly more distant position, to match the *sound perspective* with the pictorial distance.

Avoid shooting mike shadows
By careful framing you can often keep distracting mike shadows out of shot as they fall on nearby backgrounds.

Have the shot ready . . . make sure your shot is cleared.

The Cameraman and the Switcher

The *production switcher* (*vision mixer, video switching panel*) in the control room enables pictures from the various video sources to be intercut or combined as the director requires.

The switcher may be operated by the *director* himself, or by an *engineer* (the *technical director*), or a specialist operator (*switcher, vision mixer*).

The cameraman can help the switcher

If you are working in a multi-camera production with rapidly changing shots and frequent intercutting between cameras, you need to have a pretty close rapport with the person operating the switcher to avoid mutual frustration.

In a fast-moving production with few cameras, it requires a cool head to get each new shot quickly, holding it steady, sharp and well-composed in an instant, ready to be switched to the main studio output/transmission channel. If the switcher cuts before the cameraman is ready, a bad shot may be transmitted (defocused, off-subject, moving into position). If a cameraman is slow getting his shot, the action may have moved on, e.g. the person at the door has now walked through it.

Whenever a camera's picture is switched 'on-air', a red *tally light* (*cue light*) is illuminated at the front of the camera, with a corresponding indicator in the viewfinder. As the switcher cuts to the next camera, the tally light goes out and you are free to move to the next set-up.

It is particularly important to keep an eye on your tally light when your picture is being combined with another camera's (superimposition, split screen, wipe, segmented shot, chromakey treatment, inserted titles, etc.), as any unplanned camera movement will ruin the combined picture.

When your shot is being *slowly mixed* (*dissolved*) to another, take care not to alter it before the transition is over. The combined effect will be confused if you move on to the next position too early.

The switcher can help the cameraman

The switcher can in his turn assist the cameraman by giving warnings at difficult moments . . . 'Coming to 2 . . . Still on 3'. Similarly, the switcher can help by waiting for a harassed cameraman to settle his shot before taking it, when there has barely been time for a camera move.

Settle shots quickly
Get to your shot as soon as possible, particularly in a fast-moving show. Until you have settled on your shot, the switcher cannot take it.

Wait to be cleared
Do not move to your next shot until your camera has been cleared, i.e. until the switcher has cut to another camera and your tally light (cue light) has gone out.

Any moment now!
The switcher can help the cameraman by warning him that his shot is about to be switched on-air.

Shooting Remotes

The *'remote'* today covers situations as diverse as news gathering (*ENG*), location recording of inserts for studio productions, documentaries, drama, exhibitions, conferences, sports events and Grand Opera. So venues and working conditions can be extremely varied: sports stadia, industrial locations, churches, theaters, golf-courses. . . .

Life for a cameraman in a remotes (outside broadcast) unit can be very different from that of his studio colleagues. In a studio the closely-knit team is usually working within sight of each other. On location, cameras may be dispersed over a vast area, each cameraman being isolated by the sheer scale of operations.

The complexity of remote operations can vary considerably. A cameraman may be working as a self-contained single-camera unit, or linked to a small two-camera van, or part of a multi-camera crew directed from a central control truck.

The *tripod-mounted cameras* used when shooting from fixed viewpoints can be remotely located in a wide variety of positions: on a platform frame (camera tower), hydraulic platform, balcony or roof-top, or surrounded by the public on a sidewalk.

Where the camera needs to be mobile, supports include hand-held, rolling tripod, pedestal, lightweight crane, *body-mounted stabilizer (Steadicam).*

The cameraman's viewpoint

Production on location has an element of improvisation, for there is always the unpredictable. Particularly when the camera is distant from the control point, the director may be very reliant upon the cameraman using his initiative and offering up those extra unanticipated shots – from the subtlety of a gesture to the split-second drama of unexpected tragedy. Such pictures make a good show great!

Lens angles

Generally speaking, the location cameraman tends to work with narrower lens angles than in the studio. He often cannot get close enough shots any other way. Using lenses of around 20° – $^{1}/_{2}$° (compared with typical 50° – 5° in studios), camera handling on movement can be that much more difficult. Depth of field is correspondingly more limited. Particularly when shooting at extreme distances, or in windy conditions, it may be impossible to hold a steady shot without locking off the tilt/pan movements of the camera head.

Lighting conditions can change considerably on remotes, from intense sunlight needing ND filters and a small lens-aperture, to situations where the lens is working wide open to cope with low light levels.

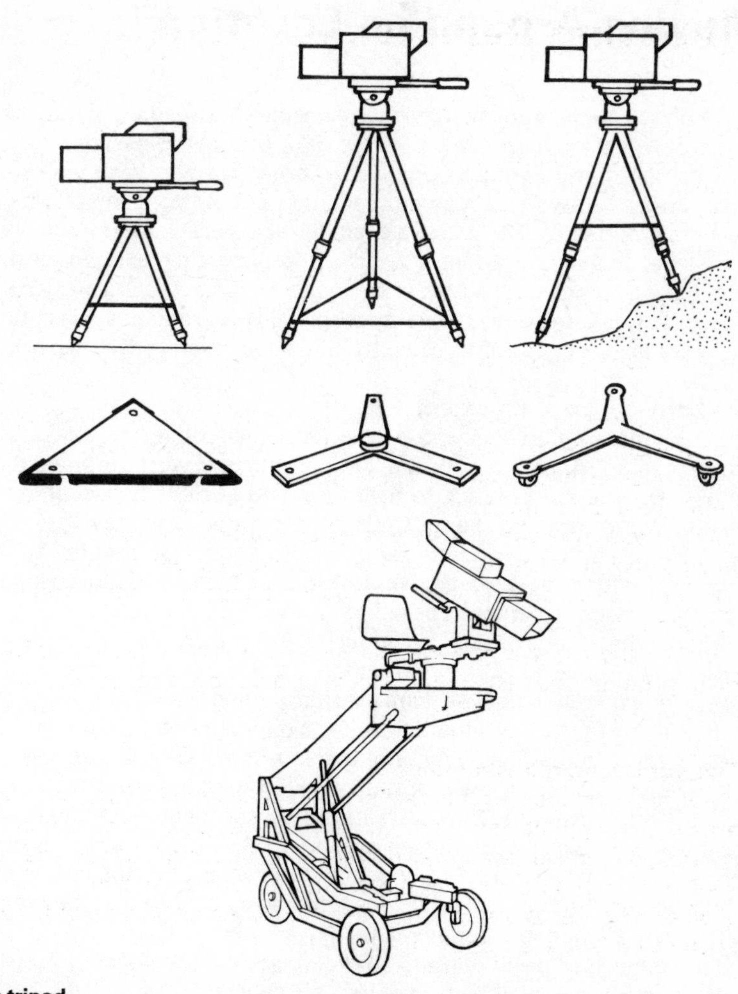

The tripod
When working on location, the tripod is an extremely adaptable camera mounting. Its legs can be linked, strapped or chained together to prevent splaying, and bottom weighted to improve stability. On rough ground, foot-spikes prevent feet from slipping. Its height is pre-set.

Tripod base
To prevent damage to floors and to avoid feet slipping, a metal triangle or a crowsfoot (spider) can be fitted. A castered (wheeled) skid may be attached to a heavy-duty tripod to form a *rolling tripod*, enabling the mounting to move around on smooth floors.

Lightweight dollies
Lightweight dollies with pneumatic tires offer increased mobility. (Studio mountings usually have solid tires for bounce-free movements, but these reveal any ground unevenness.) More complex dolly designs have pneumatic, hydraulic or electrically controlled crane arms (boom, jib).

Moving Around on Location

Single video-camera units today have considerable freedom of movement and are widely used for *ENG, EFP, EJ, ESG,* etc. (see **Glossary**).

In 'combo' form (*combination videorecorder/camera; VRC*) the camera incorporates its own videocassette recorder. But most video systems use a separate portable VTR, held in a shoulder-bag, back-pack or trolley-pack, or installed in a nearby support vehicle. The camera's maximum distance from base depends on the type of cable used – *multicore, triaxial, coaxial, fiber optics.* Lighter cables have many practical advantages – less bulky, easily transported, easier to route (can be suspended or wall-taped).

The body-supported camera

Although many *lightweight cameras* are shoulder-supported, this can be a tiring position to maintain for any length of time. Even with the aid of a *body-brace* it is not possible to hold a camera absolutely still, shot after shot for long working periods. Especially when you have to use a *narrow-angle (long focus)* lens to get a close enough shot of the subject, your own natural movements (heart beat, breathing, tired muscles) make a certain amount of unsteadiness unavoidable.

Where the scene itself is perfectly still, a wavering camera is distractingly obvious. But where there is widespread movement the viewer may well overlook camera-shake altogether. Although quite unacceptable for formal situations, a certain amount of unsteadiness may even seem to add authenticity or immediacy to a casual street interview.

If you are going to walk around while shooting, check your route beforehand, if possible, for obstacles. There are many potential hazards for the unwary, including uneven ground, cables, rugs, stairs, steps, low furniture, wet floors, etc. Some cameramen keep both their eyes open while looking through the small viewfinder eyepiece, so that they remain fully aware of their surroundings as they move.

Remember too that if you are recording sound, the noise of footsteps walking across a gravel path may be your own!

Although you can often simulate movement by zooming, remember there are always pre-focus and handling problems as the lens angle narrows, particularly with a hand-held camera.

Using a tripod

It may seem a chore to carry a tripod around, but unless shots are pretty brief with resting periods between, you will soon come to welcome the stability a tripod offers. A few words, then, about using tripods – for they can seem to develop a will of their own!

1. Tripods can become top-heavy and overbalance if their legs are not fully spread or not properly adjusted, especially on uneven ground.
2. Unless the tripod is very firmly based or bottom weighted, do not leave your camera unattended. If the pan-head is not tightly locked off, you may find the camera tilting and unbalancing the tripod.

3. The ends of tripod feet are of two kinds: spikes that grip rough ground (but can pit, scratch, or slide on other surfaces) and non-slip rubber end-pieces for smooth or easily-damaged floors.
4. Always make sure that the tripod is level and balanced. You can check whether a tripod is level by locking the tilt action and then panning around. If the tripod needs levelling, horizontals tilt down on one side (the leg that is too low).
5. You should normally fully extend all three legs of the tripod. The exception is when you deliberately shorten a leg to suit uneven ground (e.g. a staircase, rocky terrain).
6. To improve a tripod's stability and/or to prevent its feet from sinking into the ground, you can fix a special spreader or base-plate to its feet.

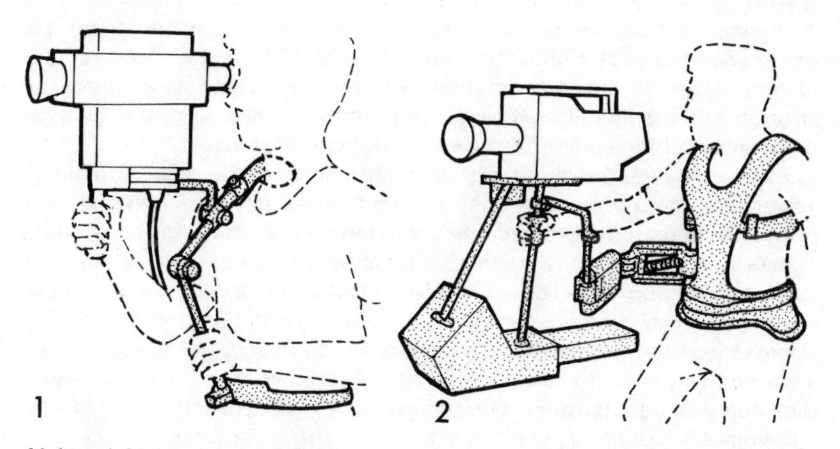

Lightweight camera supports
To increase the stability of hand-held cameras, (1) a body brace or (2) an elaborate stabilizing spring harness may be worn (Steadicam; Panaglide).

On Location – Light and Lighting

Unless you are careful, available light can produce appalling pictures. (You will find the subject discussed more fully in books listed on page 152.)

Light intensities

Where there is insufficient light to expose a shot properly, boosted video gain often produces a bolder, brighter picture (but increased picture noise). However, extra amplification of the video signal cannot improve shadow detail.

The simplest supplementary lighting (apart from a reflector board) is a hand-held or camera-mounted lamp of 100–1000 watts, powered by battery or utility supplies (mains). Sometimes a floor-stand lamp, or stronger light-bulbs in existing room fittings, may improve general illumination sufficiently or reduce overall tonal contrast.

If tones are too contrasty (e.g. sunlight and deep shadow), it may be impossible to expose both light and dark areas effectively. Where you cannot illuminate dense shadows, try to keep extreme tones out of shot.

Avoid shooting into very bright areas when using *auto-iris,* for the system will stop itself down (*iris down*) and under-expose the main subject. Closing drapes over daylit windows or changing your shooting angle to exclude such surfaces often helps. You may even have to choose a camera viewpoint to suit the existing light, rather than the best view of the subject. If you can angle the subject itself to fit the prevailing lighting, so much the better.

Light direction

Light direction affects appearance. If light comes from the camera direction (*frontal lighting*), texture and surface-modeling are reduced. Light shining across the subject (*side light*) exaggerates surface texture and contours, often creating a coarse, bisected effect. Light shining towards the camera *backlights* the subject with a surrounding rim, revealing any transparency or translucency and separating it from the background. Whether you move a light relative to your camera viewpoint, or your position relative to the light, the pictorial effect is the same.

Color temperature

The light's color quality affects the color fidelity of the picture. Orange-yellow light such as candlelight (*low color temperature*) is at the lower end of the kelvin scale, bluish light from daylight (*high color temperature*) at the upper end.

If your camera's *color balance* (red-green-blue proportions) does not match the prevailing light, the picture has a color cast. A *'white balance'* or *'auto-white'* adjusts the error – but it cannot compensate if several types of light are intermixed (e.g. daylight, tungsten, fluorescent).

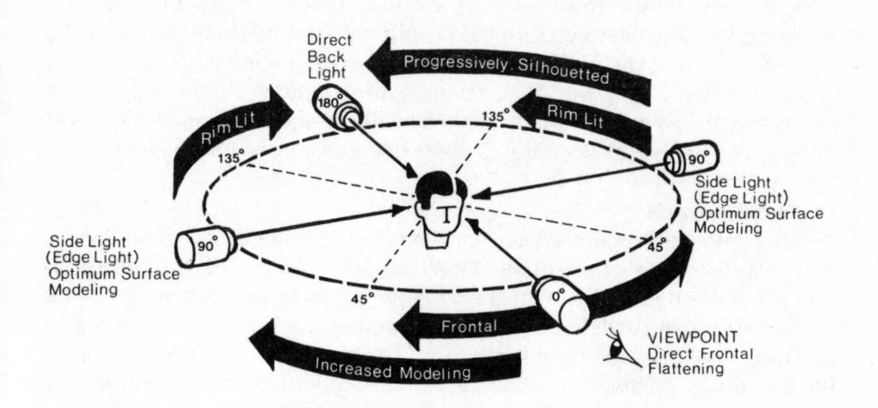

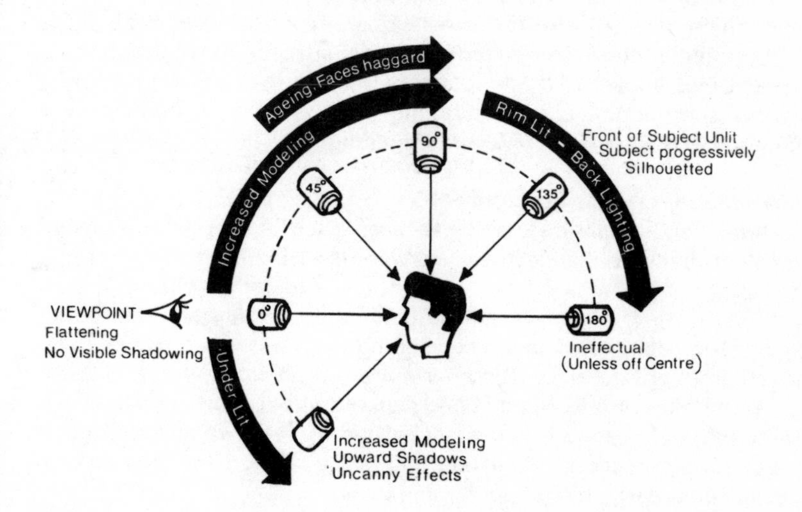

Lighting angles
The effect of light changes as its angle to the subject is altered, whether you reposition the camera relative to the light, or relocate the light. The eye here represents the camera's viewpoint.

For extreme variations in camera height.

Camera Cranes

For everyday work in the TV studio, a standard pedestal has proved to be an extremely flexible, economical mounting. One operator can move it around quite quickly, even in confined spaces, adjusting its height over a useful working range. But in larger-scale productions, where the director needs extra camera height (for an elevated viewpoint, or to see over foreground people or scenery), a *camera crane* comes into its own.

The small crane

Various manually operated dollies have been designed for film-making and adapted to television use. Small cranes have a *boom* (*crane-arm, tongue, jib*) mounted on a steerable wheeled platform. This may have a central rotatable platform (turntable) enabling the boom to be turned in any direction. The height of the boom on which the cameraman sits with the camera is adjusted by a hand-crank or counter-balance control. The lens height can be varied in shot over 1–2 m (3–7 ft) or more, the boom overhang enabling it to extend over local obstructions (e.g. a table).

The penalty one pays for this extra camera flexibility is that the crane takes up quite a lot of room, needs plenty of space to move around and requires a crew of two or three operators to assist the cameraman. While the crew control dollying and craning, the cameraman operates the camera itself, guiding the crew with finger signals.

The Academy crane (Mole crane)

This large, robust camera crane has been used in film and television studios for many years. Electrically driven, the camera at the end of its large counterbalanced boom has a height range of, for example, 60 cm to 3 m (2 to 10 ft). The camera can be panned over 180°, the main boom slewing round 360°. The camera crew comprises the cameraman, a driver (tracker) and an operator (boom-swinger) to handle the boom. An attached monitor, intercom and hand signals help to coordinate its crew.

Rigid safety precautions are essential, particularly when everyone is concentrating on action rather than their surroundings, for the boom or the base could easily overshoot its marks.

Motorized cranes

Motorized or power-operated dollies of several designs have been built for TV work. These do away with heavy manual effort, reducing crewing to just the cameraman and one other operator. Typical arrangements include floor-pedal controls for the cameraman (boom height, camera platform rotation), the dolly's direction and speed being controlled by his assistant.

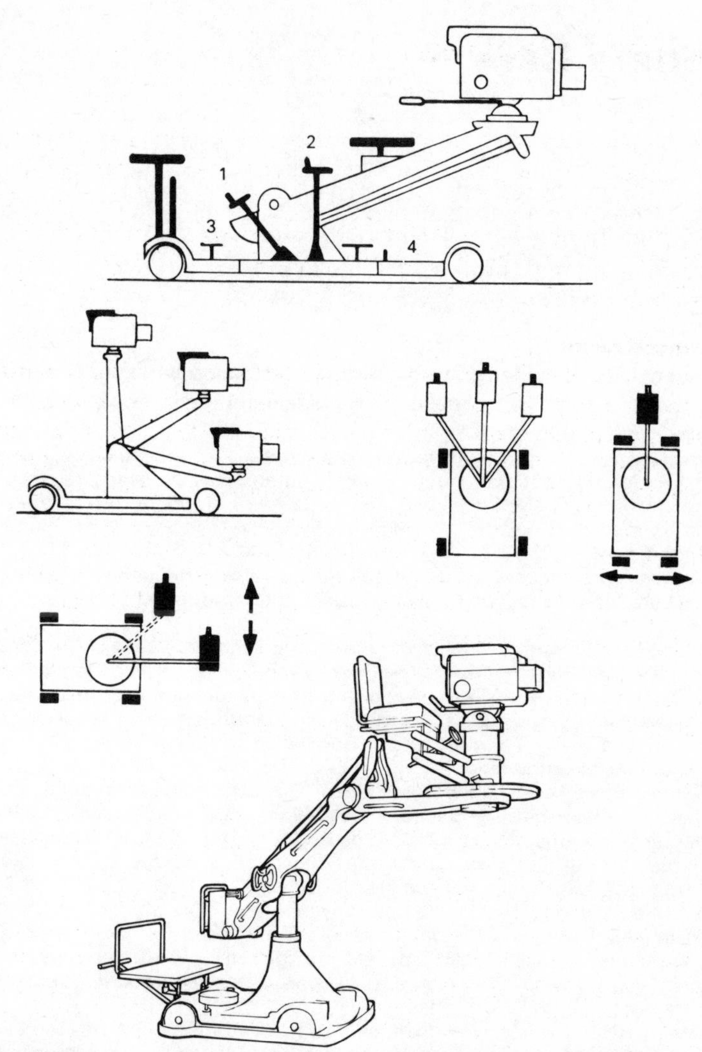

The small crane
The cameraman sits astride the crane's boom (crane-arm, jib, tongue). The crane is pushed by an operator (grips, tracker) who guides it with a tiller or wheel. A hand-crank (1) adjusts the camera height. (If a central turntable is fitted, the boom can tongue/slew sideways – hand-crank 2.) Hand-screws (3) lock dolly (and turntable) movement. Typical base 1.2×2m (4×9ft).

Crane movements
The crane can boom up/down, dolly in/out, tongue (slew, jib) left/right, in addition to camera head movements (pan, tilt). It may truck sideways.

The Academy crane
Base area 1×1.8m (3.5×6ft). Overall length 4.25m (14.5ft) maximum. Electrically driven (8kph/5mph). The centrally pivoted boom is counterbalanced (lead weights in bucket) on its height-adjustable support pillar.

Preparing for next time.

Camera Care

Having finished the day, there is always the temptation to tuck equipment away rapidly and get on with other more pressing things. But an experienced cameraman remembers that its condition at today's end will be the same at tomorrow's start. There is nothing worse than beginning the day unprepared. Throw a cable into a corner and overnight it becomes mysteriously transformed into a tangled skein!

Routine checks

Human nature being what it is, routine checks become irksome – but in the long run they pay off. So make it a regular habit to check out details after switching off the camera.

1. *Panning head.* Lock off and secure the pan/tilt movements (including securing chain if fitted). Do not tighten drag (friction) controls. Has the head been operating smoothly? (Regular cleaning/lubrication is needed.) If weighty accessories (e.g. prompter, matt box) have been removed, check that the camera-head is not now left 'back-heavy'.
2. *Pedestal column.* Lock off vertical column movement. Remember, if a prompter is removed, the column will be wrongly balanced (hard to lower) if used without one next time.
3. *Lens.* Before fixing its lens cover, look along its surface against the light. Is it clean of dust and finger marks? Lenses have a very thin surface-coating (blooming) which reduces internal reflections and improves picture contrast. This coating is easily scratched or worn, so never touch the lens surface. Use a special lens brush to dust it, or a can of compressed air. Only when these are ineffective, use lens tissues or lens-cleaning fluid.
4. *Camera head.* Is the camera head clean? Dust and grime soon build up. Have you had any mechanical or electronic trouble that now needs attention? Did the various controls operate correctly? (Focus, zoom, shot box, indicators.) Is the viewfinder OK? (Sharpness, brightness, linearity, mixed feeds, etc.) Intercom working properly? (Communal and private wires.)
5. *Camera mounting.* Have there been any problems? (Steering, elevation, motor controls, etc.) Check over the dolly's general cleanliness, including tire surfaces. Odd remnants of plastic floor-tape etc. get picked up and can cause bumpy dollying; an oil/grease streak can cause tire-slip. Check cable guards to ensure that they are not loose, too high or too low.
6. *Cable.* Having made sure that the equipment is switched off, remove the camera cable. Check that its ends are protected (capped) and store it neatly, ready for next time. Typical methods are a figure-of-eight pile on a canvas carrying sheet, or a cable drum. Finally, place dust covers over the equipment.
7. *Storage area.* Many studios simply leave cameras grouped on the studio floor at the end of the working day. If, however, scenery has to be moved in and out of the studio, it is safer to have a nearby equipment storage room, away from the general traffic.

Care with hand-held/portable cameras

It's easy to become casual and somewhat careless with lightweight cameras, especially when taking a welcome break from a long period of shooting. It is a relief to take it off and put it down. But where? Remember, your camera is really a delicate instrument. It does not like dirt, water, hot sun, moisture. The lens easily becomes coated with dust or condensation.

Grit gets into tender areas, and can scratch as it is removed. It is a good principle to fit the camera into a suitably protected container when you are not using it, or attach it to a firmly-based tripod. Today's care repays you with equipment reliability tomorrow.

Do not disconnect the camera until you have finished shooting. When you do so, fix caps (or plastic bags) over cable connections to prevent damage, damp or contamination. When winding up the cable, never pull, kink, loop or strain it, or drag it over the ground. Although it is easier to pull and wind it onto a drum like a garden hose, cables are quite vulnerable. Any damage may be internal and not obvious until you find out in use!

Batteries, too, are often taken for granted. Make sure that you have all batteries checked and re-charged after use. Don't leave them around in a half-charged condition. It shortens their life, and the running-time left in them is uncertain when you come to use them next time.

Practice Makes Perfect

There is a pride in being able to do something really well – reliably and effortlessly. But precision only comes through practice. A videotape playback of your exercises can help to reveal your progress.

Exercises can help

Here are a number of exercises to help practice your camerawork. Do them slowly at first, progressing to faster versions later. Doing several operations simultaneously can test even the most experienced cameraman.

1. Pan across a detailed wall surface, at a very slow constant speed using wide, normal, then narrow lens angles. Now slowly tilt up and down, at a constant rate.
2. Pan across a scene in which there are a number of objects at different distances from the camera, focusing on each object in turn as you pan: (a) stopping at each, (b) in one continuous pan. Try this at different speeds and distances.
3. Focus hard on a close foreground subject, then tilt or pan to a distant subject, pulling focus (refocusing) as you do so.
4. Alter your camera height while keeping the subject exactly in center frame. (Make a small mark in the viewfinder at picture center.) Go from maximum to minimum height at different speeds. Use closer and closer subjects.
5. Move the camera towards a subject (dolly in), continually refocusing for the sharpest image. Use wide, normal and narrow lens angles in turn. Try this for various distances, including tabletop objects. Now practice dollying out similarly.
6. Move the camera across the scene (truck, crab) from left to right, then from right to left, focusing on whatever comes into shot (objects at various distances). Do this again, but holding a static subject in center frame.
7. Arc round a static subject (keeping it center frame), using wide, normal and narrow lens angles.
8. Take a close shot of a nearby subject, tilt up and quickly zoom in to some distant detail. Without prefocusing, this becomes hit-or-miss. Now prefocus the distant subject and try again.
9. Repeat the effect of (5) by zooming instead.
10. Put a detailed object nearby, then check the minimum focused distance and the available depth of field, using wide, normal and narrow lens angles.
11. With the (10) setup, use various lens angles, and change the camera distance so that the screen image of the subject remains exactly the same size. (Note how other subjects' proportions change.)

Moving people

1. Follow a person moving across the scene in a mid shot, keeping him accurately framed throughout (off-center).
2. On a low camera, keep a seated person in center-frame as he stands. Try this at different distances with various lens angles, at changing speeds.
3. Follow a person moving towards and away from camera, then across camera, then on a diagonal path. Vary his distance and speed. Use wide, normal and narrow lens angles.
4. Follow someone moving away from you, dollying after him. (Different speeds, lens angles, heights.)
5. Repeat (4), with someone approaching camera as you dolly back. (Take care!)
6. Using various lens angles, check focusing problems when two subjects are at different distances from the camera. Vary their relative distances from camera.
7. Shoot two people standing side by side. Then have one person exit the shot, and reframe to suit the new situation. Holding this centered single shot, have the second person return and re-enter frame. Correct composition during the entrance.

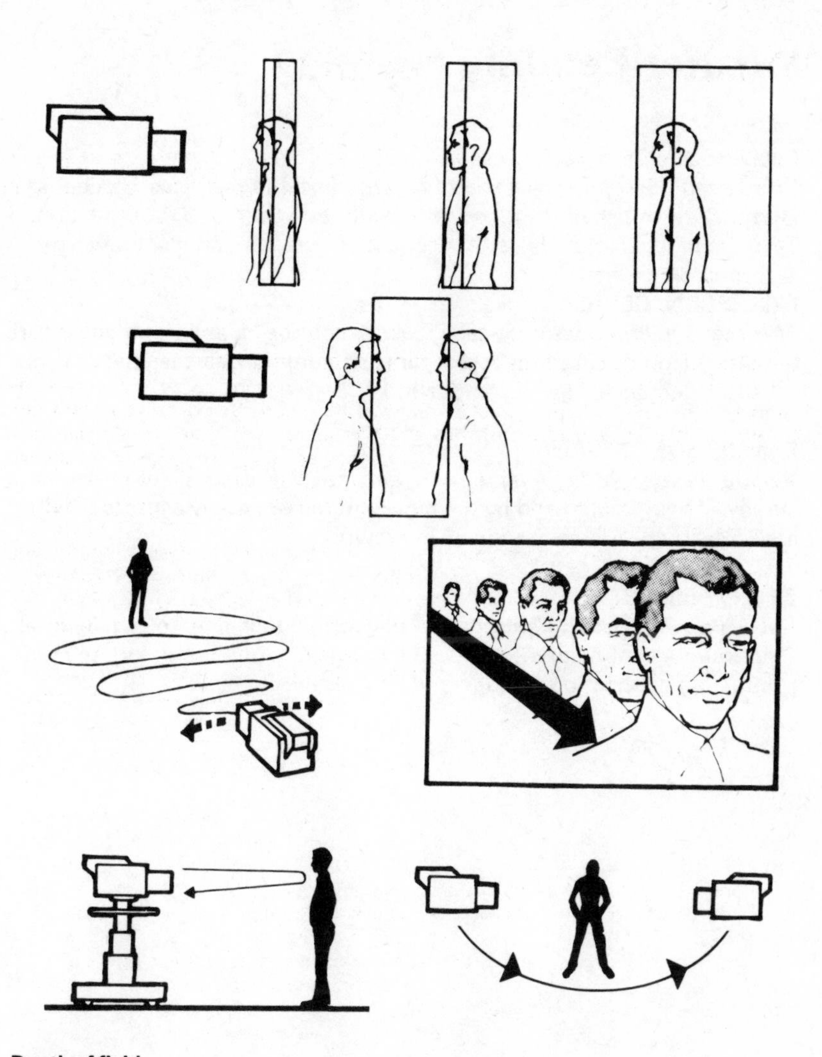

Depth of field
Check how focusing becomes increasingly critical as subjects approach the camera. Using a large lens aperture, locate the depth of field so that only the further subject is sharp. Next, split focus for equal optimum detail in both subjects. Finally, focus hard on each in turn, throwing the focus on cue.

Subject movement
Follow lateral movements, gradually coming nearer to the camera. Repeat this with narrower and narrower lens angles. Now follow a straight walk up to the camera, from the distance, at increasing speeds. Try this with wide, normal and narrow lens angles.

Camera movement
Dolly to and from the subject (maintaining sharp focus) at increasing speed, with narrower lens angles. (Take care!) Arc round a subject at different speeds while holding focus.

Further Reading

MILLERSON, GERALD:
The Technique of Television Production. Focal Press. A comprehensive study of the mechanics, techniques and aesthetics of TV production. It includes a detailed analysis of picture composition and camerawork.

MILLERSON, GERALD:
Effective TV Production. Focal Press. A concise overview of the entire process of producing video programs. It summarizes the practical and artistic essentials of good camerawork.

SAMUELSON, DAVID W.:
Motion Picture & Lighting Equipment. Focal Press. A comprehensive survey of equipment used by the film cameraman, many aspects of which are directly applicable to video camerawork.

ZETTL, HERBERT:
Television Production Handbook. Wadsworth Publishing Co. Inc. Belmont, California. A useful survey of specific television equipment, together with production applications, usage and organization.

Glossary

Action Any performance in front of the camera.
Angle of view Generally, the coverage of a lens: its horizontal (or vertical) angle. (More specifically, it is the angle subtended by the diagonal of the camera-tube target at the node of emergence of the lens.)
Aspect ratio The proportional ratio of horizontal and vertical measurement. The TV screen is standardized at a 4:3 (four by three) aspect ratio or format.

Basher A type of open reflector camera headlamp.
Blooming The bleaching-out or blocking-off to white of an area that is over-bright relative to the TV camera's tonal handling capacity. Particularly evident in specular reflections, highlights, over-lit or over-exposed areas.
Body brace A strut or framework pressing against the chest, or into a belt, to help steady a shoulder- or hand-supported camera.
Boom shot (crane shot) High-angle shot taken from a camera boom.
Burning When an excessively bright area is focused onto the camera-tube's light-sensitive surface, it may damage that region permanently (or temporarily), leaving a defacing mark on all the camera tube's subsequent pictures.

Cable-correction Electronic correction circuits that compensate for the progressive loss of 'highs' (higher video frequencies) as camera cable length is extended.
Cable drag The noise made by a camera cable as it drags across the floor surface when the dolly moves.
Cable guard A vertical metal strip attached to the base of a dolly to prevent floor cables from being trapped beneath its wheels. The guard height is usually adjustable.
Camera angle The angle the camera makes to its subject; a high-angle shot, an oblique camera angle, a low-angle shot.
Camera cable The cable attaching the camera head's circuits to the remainder of its associated electronic equipment (e.g. camera control unit). It carries video (picture signal), power, scanning waveforms, etc.
Camera card See **Shot card.**
Camera headlamp A lamp affixed to the camera head to provide local illumination close to the camera lens axis. This lamp is used as a 'fill-light' for portraiture and as a method of illuminating captions.
Camera tube The electronic device positioned within the camera head, which converts the scene focused by the lens into an electrical video signal.
Canted shot (tilted shot) A deliberately introduced visual effect in which verticals are made to lean left or right, usually to convey an unstable emotional state.

Capping-up Mechanical prevention of light reaching the camera tube by a lens cover (lens cap), obscuring the camera lens and so preventing accidental damage (burning) to the camera tube. Electrical capping-up similarly provides standby operation conditions.
Captions Any graphic, photograph, printed card, titling, etc., set up before a camera.
Cheating Changing the position of any item (person, object, scenery) already established as being in a particular place, usually to improve the composition when viewed from another camera angle.
Chinese dolly The combined effect of pullback (track back) and panning, using a tracking line angled to the subject. The result is a frontal view of a subject that progressively becomes a rear view.
Color temperature A measurement in *kelvins* of the color quality of a light source. To reproduce color accurately, the color characteristics of the video camera and the prevailing light should match.
Combination camera (Combo, CamCorder, VRC – video recording camera) A compact lightweight video camera incorporating a cassette videorecorder. Systems include 'Hawkeye' (RCA), 'BetaCam' (Sony), 'Recam' (Panasonic).
Convergence Adjustments to a color TV receiver/picture monitor to ensure the coincidence of the red, green and blue pictures.
Cover shot (protection shot) A shot embracing a wide-angle view of action, usually showing general activity from which other cameras are selecting specific changing detail.
Crabbing (trucking) Sideways movement of the camera mounting.

Defocus mix A transition between two cameras' pictures, in which the first defocuses during a mix, while the second sharpens focus on its shot.
Depth of field The range of distances over which the scene appears to be in acceptably sharp focus.
Depth of focus The extent to which the distance between a lens and a light-sensitive surface (e.g. a film) may be altered and still maintain a sharp picture.
Developing shots A sequence in which one shot is progressively modified to become another – usually through following subject or camera movement: e.g. a close shot develops as a person moves away, to become a long shot; or the camera moves around a subject to a different viewpoint.
Diascope An illuminated device containing a slide of a TV test pattern that may be attached to the front of the TV camera lens to permit electronic alignment (line-up).
Differential focusing Positioning the focused plane so that one chosen region is sharp relative to defocused surroundings.
Direction Subject directions are usually referred to relative to the TV picture. Hence 'camera left', 'camera right'.
Down stage Towards the camera.

ENG (electronic news gathering), EFP (electronic field production), EJ (electronic journalism), ESG (electronic sports gathering) Names given to various types of production using the lightweight video camera on location.

Exposure The adjustment of the reflected light values from a subject (by lighting, aperture control, filtering) to produce a particular reproduced brightness or clarity, usually relative to other subjects in the shot.

Eye line The direction in which a person is looking.

Favour To give greater prominence to one person than another in the same shot.

Filter A transparent material placed between the camera lens and the tube, with particular chosen properties that modify the appearance of the televised scene by regulating light transmission, tonal values, image clarity, etc.

Filter wheel A large, permanently fitted disc within the camera head, permitting a variety of filters to be placed behind the lens. A blanked-off section often provides the capping-up action for the camera tube.

Flare Light inter-reflecting between the various individual lenses in a lens system, resulting in reduced image contrast and spurious light patches, veiling, etc. Considerably reduced by surface coating of lenses, and by electronic flare-corrector circuits.

Floor marking Chalked, crayoned or taped marks on the floor when shots have been arranged during rehearsal, to ensure identical camera mounting positions (and furniture, scenery positions) in subsequent rehearsal and recording.

Focal length The distance between the lens system's optical center and the camera tube's target (light-sensitive surface), when focused at infinity (far distance).

Follow focus To maintain focus on a subject during movement (of subject or camera).

Foreground The part of the scene nearest the camera.

Frame jumps Changes in the frame position of a camera subject in successive camera shots, e.g. a flower vase may jump from left to right of the frame on cutting.

Framing The act of positioning a subject relative to the borders of the TV picture. Hence 'tightly framed', 'loosely framed' to indicate that subjects fill more or less of the picture area respectively.

Gamma A logarithmic measurement of reproduced tonal contrast. It relates the light input to the resulting picture tones.

Gassing The adjustment of gas pressure in certain designs of pedestal mountings, to suit the weight of the camera head and its associated accessories (prompter, matte box, etc.) and so permit freely balanced vertical movement of the pedestal column.

Group shot Shot embracing group of people, as opposed to a shot taking in only one individual, or a two- or three-shot.

Hard focus Sharply focused, i.e. the focused plane positioned for maximum clarity of the required subject.
Headroom The space between the top of a head and the upper border of the frame.
Headset (cans) Worn by camera and sound crew, this carries program sound in one earpiece and intercom (talkback) or PL (private line) information in the other. (Sometimes termed split-intercom.) A small attached microphone enables the operator to talk back to production and engineering personnel when necessary. Earphones alone are sometimes used, together with a camera microphone for talkback.
Highlight overload protection (HOP)/Anti comet-tail gun (ACT) Special camera-tube design feature that enriches the scanning beam (during line retrace/blanking periods) to discharge extreme highlights in the target's charge pattern. This prevents or reduces blocking-off of lightest tones and spurious 'comet-tails' behind moving highlights or specular reflections.
Hyperfocal plane When the camera lens is focused on its hyperfocal plane, all subjects between half that distance and infinity (the furthermost distance) will be in acceptable focus.

Intercom (general talkback) Circuits enabling the director and production team to intercommunicate with the studio crew, to issue instructions unheard by the talent or the studio microphones. *Reverse talkback* circuits (*PL,* private line) allow individual cameras/sound boom to contact the production control room.

Lens angle The field of coverage of a lens. See **Angle of view.**
Lens axis An imaginary line from the center of the lens system, towards the scene, passing through the exact middle of the picture (field of view).
Lens flare A spurious colored blob or streak resulting from a bright light or reflection shining into a lens system and being internally reflected.
Lens hood (sun shade) A round or rectangular tube fitted to the front of a lens barrel, to shield off direct light rays (light shining towards the camera). See **Lens flare.**
Light bias Special internal illumination of a camera tube to reduce smearing (trailing, beam lag) behind high-contrast subjects under low light conditions.
Line-up (a) *Lining-up a camera* is the technical process of adjusting the electrical supplies to the TV camera tube and its associated circuits for optimum picture quality. This may require special alignment chart(s) placed before the camera, or multi-camera matching on a standard subject, or may be carried out automatically by computation. (b) *Lining-up a shot* comprises positioning the camera and composing the picture for a specific effect.
Location Any place external to the studio, at which cameras are used. Hence 'on location', 'remote', 'outside broadcast'. Also general term used to designate the geographical position of a scene.

Locking off Operating a screw or lever on the panning head, which prevents all vertical or horizontal movement of the camera head.
Loose shot A shot in which the subject image size is adjusted to leave an appreciable area between it and the borders of the frame.

Matched shot A shot in which the frame position and size of a subject in one camera's picture is made to coincide with those in another.
Matte/foreground matte Mask, vignette. An outline shape, graphic, stencil, or electronically generated, enabling part of a shot to be isolated (usually for insertion into another).
Matte box A box mounted in front of the camera lens, to hold mattes (masks), gelatin filters, effects devices. It also provides an efficient sun shade/lens hood.
Matte shot (a) Shot in which part of a camera's picture is obliterated, the obscured section being filled in from another picture source. (b) Shot in which part of the camera's view is obscured by a foreground visual (painted or photographic). This foreground matte blends with the scene to appear part of it in the total picture.
Mixed feeds A switchable facility whereby a camera viewfinder may superimpose upon its own picture that of another video source, e.g. to align or juxtapose parts of their respective shots. See **Registration.**
Monitor, picture A high-grade 'TV set' (usually without a sound channel), used to check picture details (productional or technical). Can be fed from a selected video source (e.g. Camera 1's output), the studio (program) output, or switchable to any chosen video.
Motorized dolly A dolly entirely driven by electric motors (speed, craning, etc.) as opposed to manually operated dollies.

Objective camera treatment The customary productional approach, in which the camera takes the standpoint of an onlooker relative to action, with no simulated involvement.
Off stage Outside the area of the staging (setting).
On stage Towards the centre of the setting area; within the staging area.
Over-exposure Excessively bright reproduction of a subject; in an extreme case it becomes a detailless blank white area.
Overscan The extension, beyond the normal picture tube or camera tube edge limits, of the picture scanning lines. This results in a black border or edge information loss.
Over-shoulder shot Shooting past the rear of one person's head, to a scene or another person.

Panning handle (pan handle) A metal tube attached to the panning head, enabling the direction of the camera head to be controlled. Single panning handles often have an adjustable central joint. Some cameras use two handles (either side of the panning head) to guide the camera, held at arm's length.

Panning head (pan head, camera mounting head) Bolted to the top of the camera mounting, this device allows the attached camera head to pan and tilt. It is fitted with adjustable frictional braking and locking controls. Designs use friction or fluid (silicon) damping.
Parallactic movement The relative differences in displacement rates seen between objects at different distances from the camera when a camera moves through a three-dimensional scene.
Pepper's ghost A plain glass sheet placed at 45° to the lens axis to reflect directly onto a caption, so reducing inter-layer shadowing to a minimum.
Point of view shot (POV) Subjective camera treatment. A shot in which the camera viewpoint is that of a person (or object) in a scene.
Prompter A device attached to a camera to provide a performer with visual cues or script. It may range from ring-mounted cards to sophisticated electronic systems showing on a television display the image of a remotely controlled script roll (Autocue, Teleprompter).
Pull back Track (dolly) back.
Pull focus To refocus (usually rapidly) from one focused plane to another; e.g. refocusing between people at differing distances from the camera.

Reaction shot A shot showing a character's response to an event.
Registration (a) The electrical adjustment of a color TV camera to ensure the coincidence of the red, green and blue pictures. (b) The process of adjusting two or more pictures so that their combined images have a particular relationship.
Reverse angle A shot showing a subject from the reverse direction; e.g. a shot showing a person from behind is the reverse angle to a full face shot.
Reverse cuts The effect seen when a subject points in one direction in one shot, but appears to face the opposite direction in the succeeding shot.

Safe action area Area of the TV screen within which all important action must be framed to avoid cut-off.
Safe title area Area of the TV screen within which all important graphics information must be framed to avoid cut-off.
Safety harness A webbing harness to protect the cameraman on a high camera boom from falling off the mounting.
Shader An engineer/operator, remotely adjusting exposure and various video parameters for optimum picture quality (black level, color balance, video gain, gamma). Vision control operator; racks.
Shoot-off When a camera sees past the limits of a setting to the studio beyond, it is said to be shooting off.
Shot box A push-button unit that enables any pre-set angle to be selected on a zoom lens system. A line-up chart is used to adjust the system to specific angles.
Shot sheet A card or sheet attached to the camera head, showing successive positions, shots, etc. for that camera.

Single-camera techniques Production treatment in which a single camera provides all or most of the pictures during a production, all changes of viewpoint being achieved by its tracking, zooming, etc., rather than by cutting to cameras in other positions.
Size of shot The proportion of a subject or scene filling the screen.
Soft focus Focus at other than the optimum on a subject or scene. Soft focus may result from operational misjudgement, from an indistinct viewfinder, or deterioration in electronic focus of the camera tube. Sometimes soft focus is deliberately introduced to reduce detail or modeling in a subject.
Split focus The distribution of a limited depth of field so that optimum focus is achieved on the required subjects at differing distances from the camera.
Stabilizer, body mounted A special body harness incorporating stabilizing springs which provide smooth, steady camera operation, even when running, walking, climbing stairs etc. with the camera.
Sticking A camera tube condition in which an image – particularly of a very bright surface – remains visible on a picture after the subject is no longer there. (Image retention, stick-on, memory.)
Subjective camera treatment The productional use of the camera to represent a person's viewpoint, with simulated movement, etc. (See **Objective camera treatment.**) The camera may appear to take on action involvement.

Tally light A small, low-powered indicator light affixed to the top of the camera to show when that particular unit has been selected onto the 'transmission' channel by the switcher.
Thirds A rule-of-thumb concept for arranging pictorial composition, in which the picture area is divided into thirds and principal subjects arranged on these divisions.
Three-shot Shot containing three people, often isolating them from a group.
Throw focus To readjust focus from one distance to another.
Tight shot A shot in which the selected subject fills or nearly fills the screen; a close shot.
Tilt wedge A wedge-shaped attachment fixed to a panning head to enable the camera head to be tilted downwards to an abnormal degree. With it, the camera lens can take steeply tilted shots from balconies, towers, etc., of nearby action.
Tracker An operator who assists the cameraman by controlling the speed and direction of a camera dolly, while the cameraman concentrates upon camera head movement, focusing, composition, etc.
Transmission (a) The picture selected by the director and displayed on a 'transmission monitor', to be recorded or transmitted, from a series of preview monitors showing video output of each channel (camera, slide, film). Line; studio output. (b) The final version of a production, as opposed to rehearsal versions.

Traveling shot Any shot from a moving camera.
Trucking shot Dolly shot. Often used to designate extensive dolly movement, particularly when following a moving performer in a constant-size shot. Trucking (crabbing) is lateral movement of the camera mounting.
Two-shot Shot containing two people; often isolating two people from a group.

Under-exposure Excessively dark reproduction of a subject; in an extreme case it becomes a detailless dark area.
Up stage Away from the camera.

Vision control (video control) The process of continually adjusting the exposure and electronic parameters of the TV camera system to achieve optimum picture quality. 'Shader'. 'Racks'.
Vision mixer See **Switcher.**

Wall point An electrical socket attached to the studio wall, into which the plug at one end of the camera cable is fitted. Wall outlet.
Whip pan (zip pan, swish pan, blur pan) A rapid panning movement of the camera head, showing the scene clearly at start and finish of the operation, but blurring all intermediate detail.
Wrap up To finish activities.

Zoom (a) A type of lens system of variable focal length, and hence providing a variable lens angle, from narrow through normal to wide angles of coverage. May be operated by cable control, or bar on lens mounting. (b) The action of operating a zoom lens to create a visual enlarging or diminution of the central area of a picture.